more
than
just pies

RHUBARB
more than just pies

compiled by
Sandi Vitt &
Michael Hickman

Introduction by
Lois Hole

Dale H. Vitt
series editor

Published by
The University of Alberta Press
Ring House 2
Edmonton, Alberta T6G 2E1
and
Hole's
101 Bellerose Drive
St. Albert, Alberta T8N 8N8

Printed in Canada 5 4 3 2

Canadian Cataloguing in Publication Data
Vitt, Sandi, (date)
Rhubarb

Includes bibliographical references.
ISBN 0-88864-348-9

1. Cookery (Rhubarb) I. Hickman, Michael, (date) II. Vitt, Dale H.
(Dale Hadley), (date)
III. Title.
TX803.R58V57 2000 641.6'548 C99-910725-9

∞ Printed on acid-free paper.
Printed and bound in Canada by Quality Color Press, Edmonton, Alberta.

The University of Alberta Press acknowledges the financial support of the Government of Canada through the Book Publishing Industry Development Program for its publishing activities. The Press also gratefully acknowledges the support received for its program from the Canada Council for the Arts. Royalties from the sale of this book go toward supporting the Devonian Botanic Garden.

Rhubarb: More Than Just Pies is a publication for
the book trade from the University of Alberta Press.
Illustrations, including cover, by Donna McKinnon.
Book design by Gregory Brown

Contents

Preface

This book is a compilation of information and recipes for one of my favourite plants: rhubarb. Rhubarb is a hearty plant that asks for little and produces large quantities of tart, distinctive leaf stalks for use in many food dishes. It will grow happily in the back lane next to your garbage cans in poor, gravelly soil, but when treated with respect—and given some rich soil and fertilizer—it will produce more edible petioles than can normally be used by a family. It returns each year, it generally does not spread throughout your yard, and, if given a sunny location, it will live for decades. It is perfectly hardy in the cold continental climate of western Canada; in fact, it's one of those rare plants that grows better in the north than in the south. The root stocks are inexpensive to purchase, and their food value is high. In short, it's a perfect plant!

An added benefit that we rarely consider is the production of small ivory-coloured flowers and pinkish winged seeds on the tall flower stalk. These can form an attractive accent plant.

We hope you enjoy our collected recipes. We also hope you will find the collection of facts and historical accounts of this great plant interesting and useful.

Dale H. Vitt
Director,
The Devonian Botanic Garden

rhubarb flower

Acknowledgements

Buncha Ooraikul from the Department of Agricultural, Food and Nutritional Science at the University of Alberta gave us some of the most up-to-date information on the usage of rhubarb stalks and the nutritional value of rhubarb fibre. Paul Ragan at the Brooks Horticultural Research Centre provided some additional cultivation information. We appreciate their efforts very much.

We also wish to thank Maryann Rosie, who spent many hours typing and setting up the recipes. We are very grateful to the Crafters Association at the Devonian Botanic Garden, and especially Shirley Hooper, for sharing their knowledge of the plant; and also to the Friends of the Garden, who have generously supported this project through their 1997 grant program. This book would not have been possible without their support.

We would also like to thank all those who provided us with their rhubarb recipes. Enjoy!

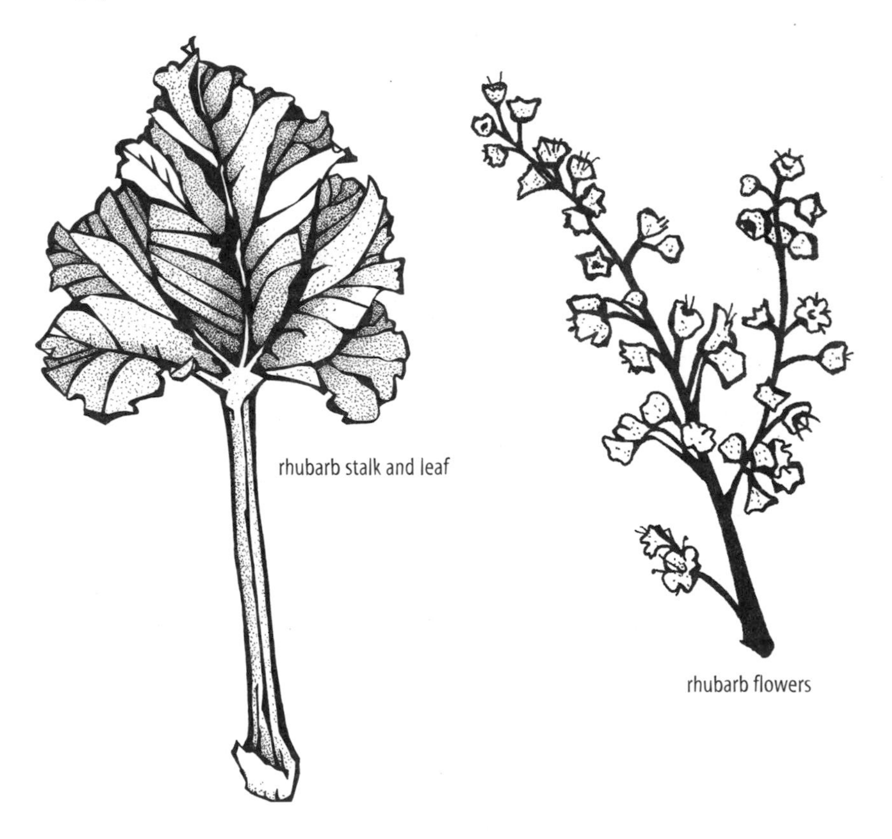

rhubarb stalk and leaf

rhubarb flowers

Introduction

If one image comes to my mind when somebody talks about rhubarb, it's probably that of a steaming rhubarb pie, fresh out of the oven. But as sweet and juicy as rhubarb pies are, they're not really exotic fare. In fact, we often take rhubarb for granted, lulled by its ubiquitous presence in pies and other favourite desserts. But over the years, I've learned that rhubarb is about more than just pies. To be sure, it's about that sharp, distinctive flavour; but to me, it's also about perseverance, simple pleasures, and memories of spring.

Rhubarb grew like a weed on the streets and in the backyards of Buchanan, the small Saskatchewan town where I grew up. Far from a nuisance, the tall crimson stalks hidden beneath those large, bristly leaves were the "Red Gold" of the Prairies. In a climate where hardiness was the ultimate garden virtue, rhubarb's tenacity made it an indispensable staple, one depended on by generations of prairie families. Fruit choices at the local grocery store were limited back then, making rhubarb's early emergence and strong flavour all the more welcome.

Since there weren't many fruits to work with, we found a number of ways to enjoy rhubarb. Pie was high on the list, of course. I remember Aunt Anna coming over several times each spring to ask, "Is there enough rhubarb for pies yet?" But I also recall coming home from school to enjoy Mom's warm rhubarb bread, a moist loaf full of rhubarb pieces, brown sugar, and invigorating spices. Mom also used to make steamed rhubarb. We would smear it on fresh bread and spread thick cream on top. Sometimes I simply dipped fresh, ruby-red stalks into a bowl of sugar and ate them raw.

But time passed, and eventually grocery stores weren't so limited in selection. Then I moved to Edmonton, and that meant even more choices. For a while, rhubarb got lost in the shuffle, but I never forgot about it entirely.

Years later, my family was operating a market-gardening business. It was a busy place, with people running everywhere, plants growing in every corner, and row upon row of carefully tended vegetables. Almost lost in the green menagerie were two long rows of rhubarb. Since rhubarb requires less care than practically any other plant, we left it pretty much to its own devices. Each year, the stalks renewed themselves predictably.

One day, a neighbour came by and asked if we had any rhubarb. "Oh, sure," I replied, "just head out back and take what you need." She went out behind the house and came back with an armload of stalks. My son Jim was working at the till. When the customer asked "how much?", Jim told her, "Oh, we don't charge for that." After all, it was a common plant and we always had too much of it. Still, the woman was thrilled to have it. Her enthusiasm made me stop and think. Sure, rhubarb was easy to grow and flourished all over the place, but did its abundance make it any less valuable? After all, we were using it for muffins, crisps—my daughter-in-law Valerie even used rhubarb as a meat tenderizer, and I've never had juicier pork tenderloins. And yet, in some ways we treated rhubarb as the black sheep of the garden, just because of its familiarity.

rhubarb seeds

more than just pies

Rhubarb • The Genus

Rhubarb belongs to the genus Rheum *of the family* Polygonaceae *(the buckwheats). This genus is comprised of some fifty species of rhizomatous, often tough and woody perennials that grow in a wide variety of habitats. Its habitats range from scrub and rocky hillsides to streambanks and wet meadows, from eastern Europe and central Asia to the Himalayas and China. Members of this genus are characteristically clump-forming plants, possessing thick, fleshy roots and large, long-stalked, basal leaves that are rounded-entire to palmately lobed. The leaves often emerge from bright-red buds and can be crimson-purple with prominent veins and midribs when young. In the late spring and early summer, these plants produce erect stems which terminate in broad, spike-like inflorescences (panicles). The inflorescences are comprised of short, axillary panicles of small, star-shaped flowers, surrounded by colourful, conspicuous bracts in some species. Small, triangular, winged, usually brown fruits follow the flowers.*

Much later, a royal visitor reinforced the point. In mid 1999, while I was serving as the Chancellor of the University of Alberta, I represented the University at a reception for the Japanese Prince and Princess Takamodo. I found out that the Prince had developed a love for rhubarb pie during his student days in Canada, and that he really missed the treat. That evening, I gathered some rhubarb from the backyard and baked up a fresh pie to present to him at the official reception at Alberta's gorgeous Government House. I was sure that no rhubarb pie had ever before enjoyed such refined settings.

Naturally, my husband Ted wasn't about to let visiting dignitaries eat any of the pie without giving it a taste test. "Hang on just a second, Lois," he said, and with a hungry gleam in his eye, he cut himself a piece, inhaled it promptly, and sent me off with his blessing.

Giving away a pie sounds like a simple gesture of goodwill, but when you're giving a gift to royalty, there are certain diplomatic rules to follow. Since I'd come up with the idea at the last minute, the presentation wasn't on the formal agenda. My pie inadvertently raised more than a few eyebrows in the provincial protocol office. At first the officials weren't going to allow it, but Edmonton Mayor Bill Smith gave me a helping hand, and together we convinced the powers-that-be that the pie wouldn't cause a diplomatic incident. Sure enough, when I presented the royals with the pie, the handsome and affable Prince was absolutely delighted. "Ted took a little piece for a taste test," I explained when the Prince noticed a missing slice. "Not such a small piece," he quipped with a grin.

I think it's interesting that the Prince saw rhubarb as a special treat, in direct contrast to our sometimes-complacent attitude about this common perennial. Familiarity may not breed contempt for rhubarb, but it does seem to encourage a certain lack of excitement and enthusiasm about it.

A Word of Caution

Oxalates are present in all parts of the rhubarb plant, especially in the leaf blades. The leaves, which contain calcium oxalate crystals, are ***poisonous****. The calcium oxalate crystals produce gastro-intestinal bleeding if eaten in large quantities; they can even cause death (although one would have to consume quite of lot of leaves as the concentration of oxalic acid is only about 5% in the leaves). There is some evidence for the presence of anthraquinone glycosides, and these, too, may be partly responsible for poisoning. The root stock should also be avoided because it also contains laxative anthroquinones. Additionally, the pollen has been known to cause allergic reactions.*

But with this book, Sandi and Michael have brought new life—and respect—to rhubarb. With dozens of fine recipes, they have reminded me of rhubarb's amazing versatility and unique, wonderful flavour. You think you know how to make rhubarb pie? Maybe so, but in this book Sandi and Michael present so many rhubarb tricks that they practically make the old favourite sing and dance. Along with Dale Vitt, they've worked for years with plants at the Devonian Botanical Garden, just outside Edmonton. Their experience really shines through in this book. Despite having access to all kinds of exotic plant life from all over the world, they obviously still care about rhubarb.

More than just pies? Absolutely. The thick patch of rhubarb that returns to my garden each year may serve as a handy source of stalks for desserts, but more importantly, it's a connection to my past. They'll love this book in Buchanan, and in the hundreds of other prairie communities where rhubarb continues to cling tenaciously to the rugged earth. Enjoy!

Lois E. Hole

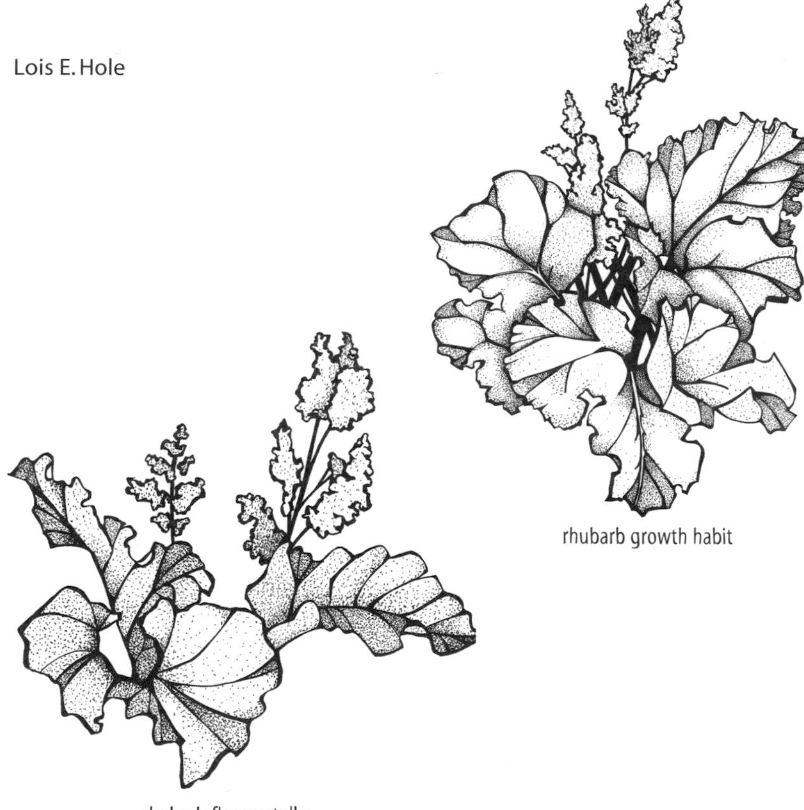

rhubarb growth habit

rhubarb flower stalks

soups

more than just pies

Nutritional Value of Rhubarb Fibre

Rhubarb stalks contain about 5% dry matter, 74% of which is dietary fibre, balanced by carbohydrates, ash, and some protein. In the 74% dietary fibre, 66% is insoluble and 8% is soluble fibre, consisting mainly of pectic substances. Rhubarb fibre is unique in that it may be processed into a dry powder that has the capacity to hold almost 20 times its own weight of water. Studies with animal and human subjects at the University of Alberta have revealed that rhubarb fibre is very effective in the reduction of cholesterol and triglycerides; it also appears to be an effective bowel regulator. These nutritional properties have been ascribed to its composition of soluble and insoluble fibre, its high water-holding capacity, and perhaps its other unique components such as minerals and organic acids. Average cholesterol reduction, mainly in the LDL portion, in hypercholesterolemic men is 10%, while average triglycerides reduction is 18%, when subjects are fed 27 grams/day of dry rhubarb. The fibre also appears to modulate body-sugar absorption, which may be beneficial in weight-control regimens. Despite the presence of oxalic acid, the fibre was not found to interfere with calcium absorption.

—Compiled by Buncha Ooraikul

Rhubarb Soup with Strawberries, Mango, and Passion Fruit

Makes 8 servings.

1	orange (zest)	1
1	lime (zest)	1
$\frac{1}{2}$ stalk	lemon grass, chopped	$\frac{1}{2}$ stalk
4 cups	rhubarb	1 L
4 cups	water	1 L
$\frac{2}{3}$ cup	sugar	160 mL
$\frac{1}{3}$ cup	grenadine	80 mL
1 tbsp.	fresh lime juice	15 mL

Place lime and orange zests and lemon grass in the center of a double-thick square of cheesecloth, gathering up the corners to form a sachet. Coarsely chop rhubarb and place in saucepan with water and sugar; bring to a boil. Remove the pan from heat, add the sachet, and cover the pan. Set aside to steep for 2 hours. Discard the sachet. Stir in grenadine and lime juice. Cover and refrigerate until chilled, about 2 hours. Garnish before serving.

Garnish

2 cups	strawberries	500 mL
1	mango	1
2	passion fruit	2
2 cups	passion fruit or mango sorbet	500 mL
	fresh mint sprigs	

Cut strawberries into slices $\frac{1}{4}$ inch (6 mm) thick, then cut slices into $\frac{1}{4}$-inch (6 mm) sticks. Peel and pit mango; cut mango into $\frac{1}{4}$-inch (6 mm) sticks. Cut passion fruit in half and scoop the seeds into a small bowl. Ladle the soup into 8 chilled shallow bowls. Place a small scoop of sorbet in the center of each bowl, and top with a sprig of mint. Garnish each bowl with strawberries, followed by criss-crossed strips of mango. With a teaspoon, drop 5 equally spaced dollops of passion fruit seeds around the perimeter of the soup.

Rhubarb Soup

Makes 8 to 10 servings.

8 cups	rhubarb	2 L
8 cups	water	2 L
1 stick	cinnamon	1 stick
2 slices	lemon	2 slices
1 ½ cups	sugar	375 mL
2 tbsp.	cornstarch	30 mL
⅓ cup	cold water	80 mL
1	egg yolk, beaten	1
½ cup	whipping cream, whipped	125 mL

Cut up rhubarb and cook in 8 cups (2 L) of water until tender. Drain liquid through sieve, discarding pulp. Place juice in saucepan, add cinnamon stick and lemon slices, and cook for 5 minutes. Add sugar. Mix cornstarch with ⅓ cup (80 mL) of cold water and stir into hot juice. Cook, stirring constantly, for 5 minutes. Remove lemon slices and cinnamon stick. Just before serving, combine beaten egg yolk with whipped cream and stir into hot soup.

Hint: *This soups is also tasty served cold with sweetened whip cream.*

Rhubarb and Strawberry Soup

Makes 4 servings.

1 cup	rhubarb	250 mL
2 cups	sliced strawberries	500 mL
½ cup	orange juice concentrate	125 mL
½ cup	water	125 mL
½ cup	sugar	125 mL
2 tsp.	sherry	10 mL
	sour cream for garnish	
	strawberries for garnish	

Put rhubarb, strawberries, orange juice, water, and sugar into saucepan. Bring to boil. Simmer until cooked. Purée in a blender. Rub through sieve to remove rhubarb fibres. Add sherry. Chill. Just before serving, place a dollop of sour cream in the center of each bowl and garnish with a strawberry slice.

Iced Tomato and Rhubarb Soup

Makes 6 servings.

4 cups	rhubarb	1 L
2 tbsp.	brown sugar	30 mL
½ cup	white sugar	125 mL
½ tsp.	ground cinnamon	2 mL
1 lb.	ripe tomatoes	454 g
1 cup	chicken stock	250 mL
1 cup	tomato juice	250 mL
1 tsp.	fresh thyme **or** ½ tsp. (2 mL) dried thyme	5 mL
1 tbsp.	chives, chopped	15 mL
2 tbsp.	fresh parsley, chopped	30 mL
	salt, to taste	
	pepper, to taste	

Cut rhubarb into ½-inch (12-mm) pieces and combine with sugars and cinnamon in a medium saucepan. Cover and cook until rhubarb begins to soften, about 8 minutes. Seed and chop tomatoes and add to rhubarb mixture. Cook 5 minutes. Stir in stock, tomato juice, and thyme. Bring to a boil, reduce heat, and simmer, uncovered, until rhubarb begins to disintegrate. Purée in food processor, cover, and chill. Sprinkle with chives and parsley, season with salt and pepper, and serve.

Did you know?

Rhubarb stalks tend to develop more red colour when the growing temperature is low, more green colour when the growing temperature is high.

Rhubarb-Cranberry Soup

Makes 4 servings.

1 cup	cranberry juice	250 mL
⅓ cup	sugar	80 mL
1½ tsp.	cornstarch	7 mL
2 sticks	cinnamon	2 sticks
pinch	nutmeg	pinch
2 cups	rhubarb, chopped	500 mL

Combine ½ of cranberry juice with next 4 ingredients. Heat and stir until mixture thickens. Add rhubarb and continue cooking slowly for 10 minutes, until rhubarb is tender. Add remaining cranberry juice. Cover and chill. Remove cinnamon sticks and serve cold.

Coconut Curry Soup

Makes 6 servings.

½ lb.	soba noodles	225 g
2 tsp.	cooking oil	10 mL
1 cup	rhubarb	250 mL
1 cup	water	250 mL
1 stalk	lemon grass	1 stalk
2 2 tsp.	red hot chilies **or** Chinese chili sauce	2 10 mL
3 cups	coconut milk	750 mL
2 cups	chicken broth	500 mL
4 slices	fresh ginger	4 slices
3 tbsp.	fish sauce	45 mL
½ tbsp.	curry powder	8 mL
2 tbsp.	lime juice	30 mL
4	chicken breasts, skinless, boneless halves, divided	4
1 cup	button mushrooms	250 mL
	salt, to taste	
¼ cup	cilantro sprigs	60 mL

Bring 5 quarts (4.7 L) of water to a rapid boil. Add the noodles; cook until tender (about 5 minutes). Drain, rinse with cold water, and drain again. Stir in the cooking oil and set aside. Combine rhubarb and 1 cup (250 mL) water and bring to boil; lower heat and cook slowly for 10 minutes. Strain the rhubarb pulp, discard, and save the liquid. Cut stalk of lemon grass into 1-inch (2.5 cm) pieces and sliver the hot chilies. In a 3-quart saucepan, combine the reserved rhubarb liquid, lemon grass, chilies or chili sauce, coconut milk, chicken broth, ginger, fish sauce, curry powder, lime juice, chicken breasts, mushrooms, and salt. Bring the soup to a simmer, stirring occasionally, for 20 minutes. Garnish with cilantro sprigs and serve.

Rosy Soup

Makes 6 servings.

1 ½ cups	rhubarb, chopped	375 mL
1 ½ cups	fresh strawberries	375 mL
1 stick	cinnamon	1 stick
1 cup	sugar	250 mL
⅓ cup	red wine	80 mL
⅓ cup	club soda	80 mL
½ cup	strawberries, sliced	125 mL

Place rhubarb, strawberries, cinnamon, and sugar in saucepan; add water to cover. Bring to a boil. Reduce heat and simmer 5 to 7 minutes, until rhubarb is tender. Remove from heat and force mixture through a sieve into a large bowl. Add wine and soda. Chill. Garnish with sliced strawberries and serve.

Did you know?

The only major disease to affect rhubarb is redleaf (also known as leaf spot). Redleaf starts with the appearance of greenish yellow areas on the leaves. These patches later turn white in the centre, with reddish edges. The leaves begin to droop, and the plant loses vigour. If left untreated, the disease will eventually kill the crown. To treat redleaf, remove and destroy all infected leaves. If symptoms continue, dig up the crown, destroy it, and start again with a healthy plant.

beverages

more than. just pies

Wines From Rhubarb

WIne was first made from rhubarb around 1840. It was as a wine that rhubarb's faculties were most trumpeted, and the Gardeners' Chronicle *(No. 19: 321) carried a recipe for it in 1843. The recipe is as follows.*

"To one pound of bruised rhubarb add one quart of cold spring water, letting the mixture stand for three days, stirring it twice a day. Then, press and strain through a sieve, and add two and one half pounds of loaf sugar for each gallon of liquor. For flavour, add a bottle of white brandy for every five gallons, and pour into barrels. The product should be left to age for at least six months, until the sweetness is off sufficiently. Finally, pour the wine into bottles."

Of course, like any recipe, it failed to satisfy everyone. In 1850, Mr. Stone of Bradford, Yorkshire, concocted a variant and patented his process. After all fermentation had ceased, he added sugar (three pounds per gallon), a technique usually used to produce effervescent wines.

However, Henry W. Livett, a surgeon living in Wells, Somerset, objected to both these processes because, when properly made, rhubarb wine has a distinctive and appealing taste that brandy ruined, and effervescent wine, according to Livett, was "indistinguishable from Champagne." For manufacture of large quantities he published a recipe in the Cottage Gardener *in 1850 (4:319-320).*

Livett's ingredients were 60 pounds of unpeeled rhubarb, 30 pounds of loaf sugar, and 4 ounces of red argol (a source of tartaric acid—an impure potassium bicarbonate found in wine casks and precipitated during fermentation of foreign wines). The rhubarb, bruised and macerated for twelve to sixteen hours, was to be left two to three days at a temperature of 56°F, then strained into a cask filled to the bung hole. The surplus "must" (about one and a half gallons) was to be stored in a jar so that it could be used to top up the cask as fermentation proceeded. Fermentation had gone far enough when the saccharometer registered forty. Then the liquor was to be fined and drawn off into a clean cask and tightly stoppered. If an effervescent wine was desired, bottle early; for a still wine, it was to be kept in the wood a year or more.

Livett's enthusiasm was shared by others, and word of rhubarb wine's progress in Britain spread quickly abroad. Although it met with a lukewarm reception in Europe, it was a hit in the United States. J.R. Mydge of Belvedere, Illinois, introduced rhubarb wine to the market in the early 1860s, and a "strawberry variety" appeared in Detroit shortly afterwards. This wine also had a discernible medicinal effect of promoting "a gentle movement of the bowel usually following its use."

Rhubarb Wine

Makes 25 750-mL bottles.

6	campden tablets, crushed	6
20 to 23 lbs.	rhubarb	9 to 10 kg
15 lbs. 12.5 lbs.	corn sugar **or** white granulated sugar	6 kg 5 kg
5 gal.	warm water	19 L
50 oz.	white grape concentrate	1.4 L
2½ tsp.	yeast nutrient	12 mL
2½ tsp.	grape tannin	12 mL
	Lalvin wine yeast	
	Isinglass finings	

Day 1. *Be sure to clean fermentor (large tub) before use! Sterilize by combining 2 of the campden tablets with some warm water and wash the fermentor thoroughly with this mixture. Cut rhubarb into 1-inch (2.5 cm) pieces and put in primary fermentor. Pour sugar over fruit (or mix sugar into rhubarb if you want). Cover fermentor with a plastic sheet and allow mixture to stand 24 to 48 hours. Add warm water and immediately strain out rhubarb. Add grape concentrate, yeast nutrient, and grape tannin, making sure to use level teaspoons. Stir well. This mixture is called the "must." Let it cool to 70–75°F (21–23°C), then sprinkle the yeast over the must.* ***Do not mix*** *the yeast into the must!*

Day 7. *Siphon (rack) the must into a 5 gal. (19 L) carboy (or smaller jugs if you are making a smaller quantity) and attach fermentation locks. Racking is most easily done with 2 carboys, moving from one to the other. The lock should be filled with a campden solution.*

Day 16. *Rack.*

Day 40. *Rack. During this last racking, shake the carboy into which you are siphoning the wine vigorously, as you are siphoning to remove any remaining carbon dioxide (CO_2). This last step will help the wine to soften and clarify faster. Add recommended finings.*

Day 47. *Bottle after the wine has cleared. Store the wine in a cool, dry place. The wine should be drinkable in 6 months; however, it will improve considerably after a year.*

Hint: *This wine is best made with young rhubarb. The colour of the wine is determined by the redness of the rhubarb used.*

Rhubarb Mead

Makes 6 750-mL bottles.

18 cups	rhubarb, chopped	4.5 L
1 package	champagne yeast	1 package
12 cups	honey	3 L
48 cups	water	12 L
1	lemon	1
1	tea bag	1

Prepare yeast as instructed on package. Dissolve honey in water. Squeeze juice from lemon and add to water/honey mixture. Stir in rhubarb. Combine this mixture with yeast and place in sterilized carboy. Let sit for 4 to 6 weeks in a dark place with even temperature. Rack to remove solids. Repeat if necessary. Bottle into sterilized wine bottles. This makes a dry, faintly pink mead or malomalt.

Fruity Rhubarb Nectar

Makes 6 cups (1.5 L).

10 cups	rhubarb, chopped	2.5 L
3 cups	water	750 mL
1 strip	orange rind	1 strip
1 strip	lemon rind	1 strip
2 cups	sugar	500 mL

In large saucepan, combine rhubarb, water, orange and lemon rind; bring to boil over high heat. Reduce heat; cover and simmer for 10 minutes or until rhubarb is soft. Strain through cheesecloth-lined strainer into clean saucepan. Stir in sugar and bring to boil. Pour into hot sterilized jars, leaving ¼-inch (6 mm) headspace. Seal with sterilized lids and process in boiling water bath for 5 minutes. Shake or stir before using.

Rhubarb Cooler: *Combine 2 parts Rhubarb Nectar with 1 part cold soda water or orange juice. Garnish each glass with strips of orange rind or orange wedge.*

Rhubarb Spritzer: *Combine 2 parts Rhubarb Nectar with 1 part each white wine and soda water.*

Rhubarb Nectar

Makes 4 cups (1 L).

8 cups	rhubarb, chopped	2 L
4 cups	water	1 L
¾ cup	sugar	180 mL
1 tbsp.	lemon juice	15 mL

Combine all ingredients and simmer 10 minutes. Strain through cheesecloth. Refrigerate. Just before serving, combine 2 parts of this rhubarb nectar with 1 part soda water.

Sparkling Ruby Punch

Makes 5 cups (1.25 L) or 10 4-oz. (120 mL) servings.

1½ cups	whiskey*	375 mL
1 cup	rhubarb nectar (use either recipe above)	250 mL
3 cups	sparkling raspberry juice	750 mL

Combine whiskey and rhubarb nectar in a large pitcher. Chill for at least 2 hours before serving. Just before serving, slowly pour in sparkling raspberry juice. Stir briefly and gently. Serve in flutes.

***American style:** use Jim Beam Kentucky Bourbon. Canadian style: use Crown Royal Rye Whiskey. Expensive style: use Jack Daniel's Tennessee Bourbon.*

Did you know?

To achieve vigorous growth, water your rhubarb well. Give each plant a thorough drink every 7 to 10 days, particularly in July, when daytime temperatures are highest. A good soaking in early fall will also help the crown to overwinter better.

Simple Rhubarb Punch

Makes 6 to 8 servings.

4 cups	rhubarb	1 L
4 cups	water	1 L
1 cup	sugar	250 mL
	sliced lemons **or** mint springs for garnish	

Combine rhubarb and water. Cook 10 minutes, until rhubarb is tender. Allow to cool. Strain and add sugar to taste. Serve iced with floating lemon slices or mint.

Rosy Rhubarb Punch

Makes 16 cups (4 L).

7 to 8 cups	rhubarb	1750 to 2000 mL
3 cups	water	750 mL
6-oz. can	frozen lemonade	180-mL can
2 6-oz. cans	water	2 180-mL cans
½ cup	sugar	125 mL
1 32-oz. bottle	ginger ale	1 L

Place rhubarb and water in large saucepan. Simmer until tender. Strain. Chill juice (you should have about 4 cups [1 L]). Add frozen lemonade, water, and sugar. Just before serving, add ginger ale.

Rhubarb Liqueur

Makes 6 cups (1.5 L).

1 cup	rhubarb, finely chopped	250 mL
½	orange, peel only	1/2
6	artichoke leaves	6
1 cup	vodka	250 mL
1 cup	sugar	250 mL
4½ cups	semi-dry white wine	1125 mL

Add rhubarb, orange peel, and artichoke leaves to vodka and seal for 5 days. Dissolve sugar in wine and add to vodka mixture. Let stand for 5 days. Strain and seal tightly. Let steep for 2 months.

cobblers, crisps, & puddings

more than just pies

Some Interesting Uses for the Rhubarb Stalk

The versatility of rhubarb stalk as an ingredient in a vast array of food products has been amply demonstrated in this book. Food scientists at the University of Alberta have developed many other interesting uses. For example, rhubarb stalks may be processed into "fruit leather" or raisin-like products. Rhubarb juice or pulp may be used as a coagulant for dairy products such as quark and high-fibre yogurt. Unique fibre strands may be separated from the rhubarb stalk and used as a texturizing component to improve the chewability or mouth feel of products such as meat patties or jerky made from mechanically deboned meats, imitation seafood products such as those made from surimi (refined fish protein), or vegetarian products made from soy protein and/or gluten.

—Compiled by Buncha Ooraikul

Strawberry-Rhubarb Crisp

Makes 6 servings.

4 cups	rhubarb, sliced	1 L
2 cups	strawberries, sliced	500 mL
⅓ cup	brown sugar, packed	80 mL
½ cup	rolled oats	125 mL
½ cup	flour	125 mL
⅔ cup	brown sugar, packed	160 mL
½ tsp.	cinnamon	2 mL
¼ cup	butter or margarine	60 mL

Preheat oven to 375°F (190°C). In a greased 8 x 8-inch (20 x 20-cm) baking dish combine rhubarb, berries, and ⅓ cup (80 mL) of brown sugar. In a bowl, combine oats, flour, ⅔ cup (160 mL) brown sugar, and cinnamon. Cut in butter until crumbly. Sprinkle crumbs over fruit. Place pan on baking sheet and bake for 30 to 40 minutes or until browned.

Rude Barb's Strawbapple Crisp

Makes 6 servings.

3 tbsp.	reduced-fat butter or margarine (not fat-free)	45 mL
¾ cup	quick-cooking rolled oats	180 mL
¼ cup	whole-wheat flour	60 mL
¼ cup	brown sugar, packed	60 mL
½ tsp.	cinnamon	2 mL
1½ cups	strawberries	375 mL
1½ cups	rhubarb	375 mL
3 cups	Granny Smith apples	750 mL
½ cup	brown sugar, packed	125 mL
2 tbsp.	orange juice	30 mL
1½ tbsp.	cornstarch	22 mL

Preheat oven to 375°F (190°C). Melt butter and combine with oats, flour, ¼ cup (60 mL) brown sugar and cinnamon in a medium bowl until mixture resembles coarse crumbs. Set aside. • Slice rhubarb and strawberries. Peel and slice apples, and combine in a large bowl with rhubarb, strawberries, ½ cup (125 mL) brown sugar, orange juice, and cornstarch. Mix well. • Spray a shallow 11 x 8-inch (28 x 20-cm) baking dish with non-stick spray. Pour in fruit and spread evenly. Sprinkle topping over fruit. Bake, uncovered, for 35 to 40 minutes, until fruit is tender and topping is golden-brown. Best served warm or at room temperature.

Rhubarb Crisp

Makes 6 servings.

5 cups	rhubarb, diced	1.25 L
dash	salt	dash
1 cup	sugar	250 mL
4 cups	dry crumbs	1 L
3 tbsp.	oil	45 mL
½ tsp.	cinnamon **or** mace	2 mL

Preheat oven to 400°F (200°C). Mix together rhubarb and salt; let stand half an hour. Add sugar and stir well. Simmer gently in saucepan for 5 minutes. Combine crumbs and oil. Heat in shallow pan in oven, stirring frequently, until golden-brown. Place a layer of rhubarb in an oiled casserole, then a layer of crisped crumbs. Repeat till dish is almost full, topping with the crumbs. Sprinkle with cinnamon or mace. Bake for 15 minutes. Serve hot with cream.

Tam's Strawberry-Rhubarb Crisp

Makes 6 servings.

2 cups	strawberries, sliced	500 mL
¾ cup	sugar	180 mL
2½ tbsp.	cornstarch	37 mL
1 tbsp.	cold water	15 mL
2 cups	rhubarb, sliced	500 mL
1 cup	flour	250 mL
1 cup	brown sugar, packed	250 mL
½ cup	rolled oats	125 mL
½ cup	butter, melted and cooled	125 mL

Preheat oven to 375°F (190°C). Combine sugar and berries. Let stand 20 minutes to produce some juice. Dissolve cornstarch in water and stir into sugared berries. Add rhubarb and spoon fruit mixture into 8 x 8-inch (20 x 20-cm) baking dish. • Combine flour, brown sugar, rolled oats, and butter, forming into large crumbs. Sprinkle crumbs over fruit mixture and bake for 35 minutes. Serve warm with vanilla ice cream or whipped cream.

Rhubarb Crisp with Bourbon Sauce

Makes 8 servings.

3/4 cup	brown sugar, packed	180 mL
3/4 cup	flour	180 mL
1 tsp.	cinnamon	5 mL
1/3 cup	butter, softened	80 mL
6 cups	rhubarb, fresh or frozen	1.5 L
1/4 to 1/3 cup	sugar	60 to 80 mL
1	lemon (zest)	1

Preheat oven to 350°F (180°C). Combine brown sugar, flour, and cinnamon. Mix in butter until crumbly; set aside. Dice rhubarb and combine with sugar and lemon zest in a 2-quart (2.2-L) casserole. Cover with crumb mixture. Bake 1 hour. Serve with bourbon sauce (recipe follows).

Bourbon Sauce

Makes 2 cups (500 mL).

3	egg yolks	3
1/3 cup	sugar	80 mL
1 cup	whipping cream	250 mL
1/3 cup	milk	80 mL
1/4 cup	bourbon	60 mL
	salt, to taste	

In top of double boiler, beat yolks and sugar until light and lemon-colored. Whisk in cream and milk. Cook over hot water (not boiling), until mixture coats a spoon. Stir in bourbon and salt to taste. Cook 5 minutes. Serve warm.

Rhubarb Raspberry Crisp

Makes 4 servings.

4 cups	rhubarb	1 L
2/3 cup	sugar	160 mL
1	orange, zest and juice	1
1 cup	flour	250 mL
1/2 cup	dark brown sugar	125 mL
1/2 tsp.	cinnamon	2 mL
8 tbsp.	cold unsalted butter	125 mL
1/4 cup	hazelnuts	60 mL
1/2 cup	rolled oats	125 mL
1/2 pint	fresh raspberries	250 mL

Preheat oven to 350°F (180°C). Combine rhubarb, sugar, and orange zest and juice in bowl. Stir; set aside. • In another bowl, combine flour, brown sugar, and cinnamon. Cut butter into small pieces and rub into flour mixture with your fingers or 2 knives, until it is well incorporated and forms large crumbs. Toast and chop hazelnuts, combine with oats, and add to flour mixture. • Turn rhubarb mixture into a 1 1/2-quart (1.5 L) baking dish, scatter raspberries evenly over surface, and cover with crumb topping. Bake about 45 minutes, until topping is crisp and juices are bubbling.

Rhubarb Crumble

Makes 4 to 6 servings.

7 cups	rhubarb	1.75 L
1/3 cup	sugar	80 mL
2/3 cup	brown sugar, packed	160 mL
1/2 cup	flour	125 mL
1/2 cup	quick-cooking rolled oats	125 mL
1/2 tsp.	ground cardamom **or** cinnamon	2 mL
1/4 tsp.	salt	1 mL
6 tbsp.	unsalted butter	90 mL

Preheat oven to 425°F (220°C). In a large bowl, toss the rhubarb with the sugar. Spread in 8 x 8-inch (20 x 20-cm) baking dish, pressing it down lightly; set aside. • In a medium bowl, combine the brown sugar, flour, oatmeal, cardamom or cinnamon, and salt. Add the butter and cut into the dry ingredients until mixture is crumbly. Spread topping evenly over the rhubarb. Bake for 15 minutes. Reduce oven to 350°F (180°C) and bake for 20 minutes or until top is brown. Serve warm or at room temperature.

Moms version – 8/9 c rhubarb, same sugar add 1/4 c grand marnier + orange peel double topping (slightly less flour) (slightly more oats) 13 x 9 pan

Rhubarb and Apple Crumble

Makes 4 to 5 servings.

4 cups	rhubarb, sliced	1000 mL
3	cooking apples, peeled, chopped	3
1/3 cup	sugar	80 mL
1 tbsp.	lemon zest	15 mL
1 tbsp.	lemon juice	15 mL
1/2 cup	orange juice	125 mL
2 tbsp.	water	30 mL
1/3 cup	raisins	80 mL
3/4 cup	dried coconut	180 mL
1/3 cup	brown sugar	80 mL
1/2 cup	flour	125 mL
1/2 tsp.	cinnamon, ground	2 mL
2 oz.	butter or margarine, cubed	60 mL
1/3 cup	almonds, slivered	80 mL

Preheat oven to 350°F (180°C). In a saucepan, combine rhubarb, apples, sugar, lemon zest, lemon juice, orange juice, and water. Bring mixture to a boil. Reduce heat and simmer 10 minutes. Add raisins. Mix well and pour into 2-quart (2-L) baking dish; set aside. • Combine coconut, brown sugar, flour, and cinnamon in mixing bowl. Work in butter until mixture is crumbly. Sprinkle mixture over rhubarb and top with slivered almonds. Bake for 15 minutes or until topping is golden-brown. Serve with ice cream, whipped cream, or whipped topping.

Apple-Rhubarb Crisp

Makes 4 servings.

2	apples, medium	2
4 cups	rhubarb, chopped	1 L
½ cup	sugar	125 mL
½ tsp.	cinnamon	2 mL
1 cup	flour	250 mL
1 cup	brown sugar, packed	250 mL
1 tsp.	cinnamon	5 mL
1 tsp.	ginger, ground	5 mL
¼ cup	unsalted butter, cold	60 mL

Preheat oven to 325°F (160°C). Grease a baking dish with at least 2 quart (2 L) capacity. Peel, core, and cut apples into ¾-inch (18-mm) pieces. Mix apples, rhubarb, sugar, and cinnamon in a large bowl. Transfer the mixture to the prepared baking dish. • Mix flour, brown sugar, cinnamon, and ginger in a large bowl. Using a pastry blender or an electric mixer on low speed, mix the butter pieces into the flour mixture until crumbs form. • Sprinkle crumbs evenly over filling. Bake until apples and rhubarb are tender when tested with a toothpick, the topping is golden-brown, and the filling just begins to bubble at the edges, about 50 minutes. Serve warm.

Strawberry-Rhubarb Crumble

Makes 6 servings.

4 cups	rhubarb, sliced	1 L
3 cups	strawberries, halved	750 mL
½ cup	sugar	125 mL
1½ tbsp.	cornstarch	22 mL
1 tbsp.	fresh lemon juice	15 mL
6 tbsp.	unsalted butter, room temperature	90 mL
⅓ cup + 1 tbsp.	almond paste, packed	95 mL
⅓ cup	sugar	80 mL
1 cup	flour	250 mL

Preheat oven to 400°F (200°C). Mix rhubarb, strawberries, 1/2 cup (125 mL) sugar, cornstarch, and lemon juice in a large bowl. Toss well to coat. Divide mixture equally among 6 1¼-cup (310-mL) custard cups. • Mix butter, almond paste, and remaining ⅓ cup (80 mL) sugar in food processor until well blended. Transfer butter mixture to medium bowl. Add flour. Using fingertips, work flour into butter mixture until moist crumbs form. • Sprinkle crumbs over rhubarb and strawberries, dividing equally. Place custard cups on baking sheet. Bake until filling bubbles and topping is golden, about 30 minutes. Serve hot.

Rhubarb Pinwheel Cobbler

Makes 8 servings.

1½ cups	sugar	375 mL
2½ tbsp.	cornstarch	37 mL
1	lemon	1
¼ cup	water	60 mL
2 tbsp.	butter, melted	30 mL
6 cups	rhubarb, diced	750 mL
1½ cups	biscuit mix	375 mL
1 tbsp.	sugar	15 mL
⅓ cup	milk	80 mL
2 tbsp.	oil	30 mL
3 tbsp.	sugar	45 mL
	butter, to taste	

Preheat oven to 450°F (230°C). Prepare lemon so that the juice can be used in fruit mixture and the zest can be used in the pinwheels. Mix sugar and cornstarch in saucepan. Add juice from lemon and water. Bring to boil, dissolving sugar. Add butter and unpeeled rhubarb. Pour into 2½-quart (2.5-L) baking dish. Set into heated oven while preparing pinwheels. • To make pinwheels, combine biscuit mix and 1 tablespoon of sugar. Stir oil into milk; add to dry ingredients. Mix lightly. Roll into rectangle ½ inch (13 mm) thick. Spread with butter. Mix remaining 3 tablespoons (45 mL) sugar with lemon zest and sprinkle half of this mixture over dough. Roll jelly-roll fashion. Cut into 8 slices and place on rhubarb in baking dish. Sprinkle with remaining zest and sugar. Bake for 25 minutes.

Rhubarb-Strawberry Cobbler

Makes 8 servings.

4 cups	rhubarb	1 L
½ cup	sugar	125 mL
1 tbsp.	cornstarch	15 mL
¼ cup	water	60 mL
2 cups	strawberries, quartered	500 mL
1½ cups	flour	375 mL
1½ tsp.	baking powder	7 mL
½ tsp.	baking soda	2 mL
¼ tsp.	salt	1 mL
¼ tsp.	cinnamon	1 mL
⅛ tsp.	nutmeg	0.5 mL
4 tbsp.	sugar	60 mL
4 tbsp.	butter or margarine	60 mL
¾ cup + 1 tbsp.	whipping cream	195 mL
1 tsp.	sugar	5 mL

Preheat oven to 400°F (200°C). In a 3-quart (3-L) saucepan, heat rhubarb and sugar to boiling over high heat, stirring constantly. Reduce heat to medium-low and continue cooking until rhubarb is tender, about 8 minutes. In a cup, mix cornstarch with water. Stir cornstarch mixture and strawberries into rhubarb mixture; cook 2 minutes until mixture thickens slightly. Remove from heat; set aside. • In a bowl, mix flour, baking powder, baking soda, salt, cinnamon, nutmeg, and sugar. Cut in butter or margarine until mixture resembles coarse crumbs. Add ¾ cup (180 mL) whipping cream; stir quickly until mixture forms a soft dough that pulls away from side of bowl. Turn dough onto lightly floured surface; knead 6 to 8 strokes to mix thoroughly. With floured rolling pin, roll dough ½ inch (12 mm) thick. With floured 3-inch (7.5-cm) star-shaped cookie-cutter, cut out as many biscuits as possible. Re-roll and cut again to make 8 biscuits in all. • Reheat rhubarb filling until hot; pour into shallow 2-quart (2-L) casserole or 7 x 11-inch (17.5 x 28-cm) glass baking dish. Place biscuits on top of rhubarb. Brush biscuits with 1 tbsp. (15 mL) cream and sprinkle with 1 tsp. (5 mL) sugar. Place sheets of foil under baking dish; crimp edges to form rim to catch any drips during baking. Bake 20 minutes or until biscuits are golden-brown and rhubarb filling is bubbly. Cool about 15 minutes. Serve warm.

Rhubarb-Tofu Cobbler

Makes 6 servings.

4 cups	fresh rhubarb, cubed	1 L
1½ cups	strawberries, sliced	375 mL
¼ cup	water	60 mL
¾ cup	sugar	180 mL
3 tbsp.	cornstarch	45 mL
10 oz.	almond dessert tofu	300 g
1	egg	1
1¼ cups	flour	310 mL
1 tbsp.	baking powder	15 mL
¼ tsp.	salt	1 mL
¼ cup	butter or margarine	60 mL
¾ cup	rolled oats	180 mL
½ cup	milk	125 mL
2	eggs	2
2 tbsp.	brown sugar	30 mL

Preheat oven to 375°F (190°C). Have a deep 9 x 13-inch (220.5 x 32.5 cm/2.5 L) glass baking dish set aside. Put rhubarb and strawberries into a large saucepan with water. In a small bowl, blend sugar with cornstarch; stir into fruit. Bring to a boil, stirring constantly, then reduce heat and let simmer for a couple of minutes. Drain tofu and spoon it into a medium bowl. Add egg. Beat with a hand beater until smooth. Pour tofu mixture into the simmering rhubarb mixture. Cook for 1 minute more. Remove hot rhubarb mixture from heat and pour into the glass baking dish; set aside. • In a large bowl, blend flour with baking powder and salt. Cut in butter or margarine until flour mixture becomes grainy. Stir in rolled oats. In another bowl, beat milk with 2 eggs and add to the dry ingredients. Stir just enough to make a moist coarse batter. • Spoon batter over the rhubarb mixture in the baking dish. Sprinkle brown sugar over the batter and bake for 30 minutes. Serve hot with ice cream.

Rhubarb Cobbler

Makes 6 to 8 servings.

6 cups	rhubarb	1.5 L
1 cup	sugar	250 mL
3 tbsp.	cornstarch	45 mL
3 tbsp.	water	45 mL
1½ cups	flour	375 mL
1 tbsp.	baking powder	15 mL
1 tbsp.	sugar	15 mL
½ tsp.	salt	2 mL
⅓ cup	shortening	80 mL
½ cup	milk	125 mL
1	egg, beaten	1
1 tbsp.	sugar	15 mL

Preheat oven to 400°F (200°C). Cut rhubarb into 1-inch (2.5-cm) pieces and place in a large saucepan. Combine sugar and cornstarch and stir into rhubarb; add water. Over low heat, cook rhubarb until juice flows. Bring to a boil 1 minute, stirring occasionally. Pour into 8-inch (20 cm/2L) baking dish. • In a bowl, combine flour, baking powder, sugar, and salt. Cut in shortening. Combine milk and egg, then add to flour mixture; stir until flour is moistened. • Drop small spoonfuls of dough into rhubarb. Sprinkle with 1 tablespoon (15 mL) sugar. Bake until golden-brown, about 25 to 30 minutes.

Scalloped Rhubarb

Makes 6 to 8 servings.

½ cup	butter or margarine	125 mL
1 cup	bread cubes	250 mL
3 cups	rhubarb, chopped	750 mL
¾ cup	sugar	180 mL

Preheat oven to 325°F (160°C). Melt butter or margarine in 1½-quart (1.5-L) baking dish. Stir in bread cubes. Cut rhubarb into small pieces and add with sugar. Bake for about 30 minutes.

Rhubarb-Peach Cobbler

Makes 8 servings.

1½ cups	flour	375 mL
3 tbsp.	sugar	45 mL
2 tsp.	baking powder	10 mL
¼ tsp.	baking soda	1 mL
½ tsp.	salt	2 mL
4 tbsp.	butter or margarine	60 mL
1 cup	plain yogurt	250 mL
1 tsp.	vanilla	5 mL
29-oz. can	peaches, sliced	770-mL can
¾ cup	sugar	180 mL
1 tsp.	cornstarch	5 mL
¾ tsp.	ground cinnamon	4 mL
⅛ tsp.	salt	0.5 mL
3 cups	rhubarb	750 mL
1 tsp.	vanilla	5 mL

Preheat oven to 400°F (200°C). Combine flour, 3 tbsp. (45 mL) sugar, baking powder, baking soda, and salt. With pastry blender or 2 knives, cut in butter until mixture resembles coarse crumbs. • Mix yogurt and 1 tsp. (5 mL) vanilla together; set aside. • Drain peaches, reserving juice. In saucepan, combine ¾ cup (180 mL) sugar, cornstarch, cinnamon, and salt. Measure reserved peach juice and add enough water to make 1⅓ cups (330 mL); stir into saucepan mixture. Add rhubarb; cook over medium heat, stirring until mixture starts to boil; continue cooking 2 minutes. Add peaches and 1 tsp. (5 mL) vanilla; return to boil. Pour hot fruit mixture into 3-quart (3-L) baking dish. • Blend yogurt mixture with flour mixture until flour is moistened. Drop dough by large spoonfuls onto hot fruit. Bake 30 minutes. Serve warm with whipped cream.

Rhubarb Charlotte

Makes 6 to 8 servings.

8 cups	rhubarb, finely chopped	2 L
1½ cups	sugar	375 mL
1½ tsp.	lemon juice	7 mL
3 tbsp.	butter or margarine, melted	45 mL
6 slices	thin-sliced white bread, crusts removed	6 slices
	brown sugar, to taste	

Preheat oven to 375°F (190°C). Combine rhubarb with sugar and lemon juice. Simmer for 5 minutes, until rhubarb is soft. • Lightly grease small charlotte mould or soufflé dish and dust with some additional sugar. Cut triangles of bread to cover bottom of container. Dip each triangle into melted butter or margarine and line bottom of container with slightly overlapping pieces. Line sides of container with more slices of buttered bread. • Fill mould with rhubarb mixture. Cover with overlapping slices of buttered bread. Sprinkle with brown sugar and bake for 30 to 35 minutes, or until crust is golden. Remove from oven and let stand for 10 minutes. Invert mould onto a heated serving dish. Serve with whipped cream, if desired.

Tapioca-Rhubarb Pudding

Makes 4 servings.

3 cups	rhubarb	750 mL
2 cups	water	500 mL
½ cup	quick-cooking tapioca	125 mL
¼ tsp.	salt	1 mL
1¾ cups	sugar	430 mL
2 tsp.	orange zest	10 mL

Cut rhubarb into 1-inch (2.5-cm) pieces. Bring water to a boil in a saucepan; add tapioca and salt. Cook over low heat, stirring until thickened, about 5 minutes. Add rhubarb and cook 10 more minutes, stirring occasionally. Add sugar and orange zest; stir until sugar is dissolved. Cool. Serve with whipped cream.

Rhubarb-Blueberry Pudding

Makes 4 to 6 servings.

2 cups	rhubarb, finely chopped	500 mL
1½ cups	blueberries	375 mL
3 tbsp.	butter, softened	45 mL
¾ cup	sugar	180 mL
¾ cup	corn flakes, coarsely crushed	180 mL
¾ cup	flour	180 mL
1 tsp.	baking powder	5 mL
¼ tsp.	cinnamon	1 mL
½ cup	milk	125 mL
1 tbsp.	sugar	15 mL
1 tbsp.	cornstarch	15 mL
⅛ tsp.	salt	0.5 mL
½ cup	orange juice	125 mL

Preheat oven to 350°F (180°C). Grease an 8 x 8-inch (20 x 20-cm) glass baking dish and set aside. Mix rhubarb and blueberries together and spread evenly on bottom of glass pan. • In a medium bowl, cream butter and ¾ cup (180 mL) sugar; stir in crushed corn flakes. In another bowl, blend flour, baking powder, and cinnamon together. Add to corn-flake mixture alternately with milk. Stir batter just to moisten. Spoon over rhubarb-blueberry mixture and spread evenly. • In a small bowl, mix 1 tbsp. (15 mL) sugar with cornstarch and salt. Sprinkle this mixture evenly over the batter layer. Heat orange juice for 1 minute in the microwave. Gently pour over the batter layer. Place in oven and bake until fruit begins to bubble, about 40 to 45 minutes. Serve with whipped cream or whipped topping.

Rhubarb-Tapioca Pudding

Makes 16 servings.

4	eggs, separated	4
1	whole egg	1
5 tbsp.	tapioca	75 mL
2½ cups	sugar	625 mL
1 cup	water	250 mL
2 tsp.	vanilla	10 mL
½ tsp.	cinnamon	2 mL
8 cups	rhubarb, chopped	2 L
	butter or margarine	
5 tbsp.	sugar	75 mL
¼ tsp.	cinnamon	1 mL

Preheat oven to 350°F (180 °C). Beat the egg yolks with the whole egg; reserve 4 whites. Add tapioca, sugar, water, vanilla, and ½ tsp. (2 mL) cinnamon. Stir and pour over rhubarb. Mix and let stand 15 minutes, stirring occasionally. Pour into large buttered pudding pan. Dot with a little butter. Bake for about one hour. • Make meringue by beating the 4 egg whites. Combine sugar and ¼ tsp. (1 mL) cinnamon and gradually fold into egg whites. Spread over pudding. Return to oven and brown lightly.

Spring Fruit Tapioca

Makes 4 to 6 servings.

¼ cup	tapioca	60 mL
1 to 1½ cups	sugar, according to taste	250 to 375 mL
½ tsp.	salt	2 mL
2½ cups	rhubarb, diced	625 mL
2½ cups	water	625 mL
1 cup	pineapple, crushed	250 mL

Combine tapioca, sugar, salt, rhubarb, and water in saucepan. Cook over medium heat until mixture boils, stirring until rhubarb is tender and sauce thickens. Remove from heat; cool. (Mixture thickens as it cools.) Add pineapple with juice; chill. Serve plain in sauce dishes, or serve in sherbet glasses topped with whipped cream. Also good as sauce over ice cream.

Hot Rhubarb Pudding

Makes 8 servings.

4	eggs, beaten	4
2 cups	sugar	500 mL
2 tsp.	vanilla	10 mL
½ cup	unbleached flour	125 mL
8 cups	rhubarb, chopped	2 L

Preheat oven to 350°F (180°C). Whisk together eggs, sugar, vanilla, and flour. Stir in rhubarb. Grease a 3-quart (3-L) baking dish and pour in the rhubarb mixture. Bake 40 minutes or until the pudding is firm. Serve warm with whipped cream.

Rhubarb Cream Pudding

Makes 6 servings.

4 cups	rhubarb, in 1-inch (25-mm) pieces	1 L
1 cup	water	250 mL
1 cup	sugar	250 mL
1 tsp.	lemon zest	15 mL
2 envelopes	unflavoured gelatin	2 envelopes
½ cup	cool water	125 mL
½ cup	heavy cream	125 mL

Combine rhubarb, 1 cup (250 mL) of water, sugar, and lemon zest in a saucepan. Cover and bring to a boil, then lower heat and simmer until tender. Sprinkle gelatin over the 1/2 cup (125 mL) cool water in a small bowl. Let stand 5 minutes to soften. Stir into hot mixture. Cook 5 minutes, mashing rhubarb. Pour into bowl. Chill until mixture will hold its shape softly when spooned. Beat cream in a small bowl until stiff. Fold into rhubarb mixture until no white streaks can be seen. Spoon into large serving dish or individual dishes. Chill 4 hours, until set.

Citrus Sponge Pudding with Rhubarb Sauce

Makes 4 servings.

2 tbsp.	unsalted butter, soft	30 mL
½ cup plus 1 tbsp.	sugar	140 mL
pinch	salt	pinch
1 tsp.	lemon zest	5 mL
1 tsp.	lime zest	5 mL
2	egg yolks	2
3 tbsp.	flour	45 mL
1½ tbsp.	fresh lime juice	22 mL
1½ tbsp.	fresh lemon juice	22 mL
¾ cup plus 1 tbsp.	milk	195 mL
3	egg whites	3

Preheat oven to 325°F (160°C). Grease 4 6-oz. (180-mL) ramekins; coat with sugar. At low speed of an electric mixer, combine butter, sugar, and salt. Mix on medium speed until crumbly. Add zests and yolks. Add flour in small amounts alternating with juices and milk. Beat egg whites until stiff. Fold into egg-yolk mixture. Ladle into ramekins; place in baking pan that has a cloth in the bottom. Fill pan halfway with hot water. Bake until puddings have set (tops may crack), about 25 minutes. Cool puddings on a wire rack 30 minutes. Invert onto plates and serve chilled or at room temperature with rhubarb sauce.

Rhubarb Sauce

1½ cups	fresh or frozen rhubarb	375 mL
3 tbsp.	sugar	45 mL
1 tbsp.	water	15 mL

Chop rhubarb into small pieces. Add to sugar and water in small skillet; cook over medium-high heat until tender, about 4 minutes. Cool.

pies

more than just pies

Species Cultivated

Among the numerous species, cultivars, and hybrids of rhubarb are some grown for their large, imposing basal leaves and tall, attractive flower panicles: for example, Rheum *"Ace of Hearts",* R. alexandrae, R. emodi, R. nobile *(Sikkim rhubarb),* R. palmatum *(Chinese rhubarb, Turkey rhubarb, or East Indian rhubarb), and* R. palmatum *"Atrosanguineum". There are also varieties grown for their medicinal properties—such as* R. officinale *(Chinese rhubarb) and* R. coreanum*—and many cultivars and hybrids are grown strictly for their edible petioles—such as* R. *x* cultorum, R. ribes, R. spiciforme, *and* R. officinale. *There is also monk's rhubarb, or garden patience, which belongs to the genus* Rumex (R. alpinus) *but was used for medicinal purposes much in the same manner as* Rheum *x* cultorum.

Blink-of-an-Eye Rhubarb Pie

Makes 6 servings.

1 cup	rhubarb, diced	250 mL
½ cup	flour	125 mL
1 tsp.	baking powder	5 mL
¾ cup	sugar	180 mL
½ cup	pecans or walnuts, chopped	125 mL
1 tsp.	vanilla	5 mL
1	egg, slightly beaten	1

Preheat oven to 350°F (180°C). Place rhubarb in a medium-size bowl. Sift flour and baking powder over top. Add sugar, nuts, vanilla, and egg. Mix all together and spread in a buttered, 9-inch (23-cm) pie plate. Bake 25 to 30 minutes.

Rhubarb Torte

Makes 8 servings.

1 cup	flour	250 mL
8 tbsp.	unsalted butter	120 mL
5 tbsp.	icing sugar	75 mL
2	eggs, well beaten	2
1/4 tsp.	salt	1 mL
3/4 tsp.	baking powder	3 mL
1 tsp.	vanilla	5 mL
1 tsp.	lemon zest	5 mL
3 cups	fresh rhubarb, diced	750 mL

Preheat oven to 350°F (180°C). Cream together the flour, butter, and icing sugar. Press into the bottom of a 9-inch (22.5-cm) glass pie pan and bake for 10 to 12 minutes. Combine eggs, salt, baking powder, vanilla, lemon zest, and rhubarb, and spread over the crust. Bake for 30 minutes more, or until centre has set. Serve warm, topped off with a spoonful of whipped cream.

Upside-Down Rhubarb Torte

Makes 8 servings.

2 tbsp.	butter or margarine, melted	30 mL
1 cup	brown sugar, packed	250 mL
	red food colouring (optional)	
3 1/2 cups	rhubarb, diced	875 mL
1 cup	flour	250 mL
1 1/2 tsp.	baking powder	7 mL
1/4 tsp.	salt	1 mL
2	eggs, separated	2
1 cup	sugar	250 mL
1 tsp.	vanilla	5 mL
1/2 cup	hot water	125 mL

Preheat oven to 350°F (180°C). Combine butter, brown sugar, and, if desired, a few drops of red food colouring in a round 10-inch (25-cm) ovenproof skillet; top with rhubarb. • In a bowl, sift together flour, baking powder, and salt. In another bowl, beat egg yolks until thickened; add sugar gradually and beat well. Add vanilla to water. Add dry ingredients alternately with vanilla-water to yolk mixture, mixing just enough after each addition to keep the batter smooth. Beat egg whites to stiff peaks; fold into batter. • Pour batter over rhubarb mixture. Bake for about 45 minutes. Invert immediately. Serve with whipped cream, if desired.

Rhubarb-Raspberry Pie

Makes 8 or 9 servings.

1⅓ cups	sugar	330 mL
¼ cup	quick-cooking tapioca	60 mL
4 cups	rhubarb, diced	1 L
1 cup	raspberries	250 mL
2 tbsp.	lemon juice	30 mL
2¼ cups	flour	560 mL
¼ tsp.	salt	1 mL
¾ cup	butter or margarine	180 mL
2 tbsp.	raspberry vinegar	30 mL
3 to 5 tbsp.	cold water	45 to 75 mL

Preheat oven to 400°F (200°C). In a large bowl, stir together sugar and tapioca. Add rhubarb, raspberries, and lemon juice; mix gently but thoroughly. Let stand 15 to 60 minutes to soften tapioca; mix gently several times. • In another bowl, combine flour and salt. Cut in butter until fine crumbs form. Sprinkle raspberry vinegar and cold water over crumbs. Stir with fork until dough holds together. Divide dough in half and pat each portion into a smooth, flat round. (You can make pastry up to 3 days early by sealing in plastic wrap and keeping in refrigerator.) • On a lightly floured board, roll half the pastry into a 12-inch (30-cm) round; ease into a 9-inch (23-cm) pie pan. Fill with rhubarb mixture. Roll remaining pastry into a 10-inch (25-cm) diameter square. With a knife, cut into ½-inch (12-mm) strips. Arrange strips on top of pie in lattice pattern; trim off as they overlap rim. Fold bottom crust over lattice; flute to seal. Set pie in a foil-lined 10 x 15-inch (25 x 38-cm) pan (pie bubbles over as it cooks). Bake on the lowest oven rack until pastry is golden-brown and filling is bubbly, 40 to 50 minutes. If rim begins to darken excessively before filling bubbles, cover with foil. Cut into wedges and serve warm or at room temperature.

Hint: *If making ahead, let pie cool completely, then cover loosely and store at room temperature for up to a day.*

Rhubarb Pie

Makes 8 servings.

1	egg white	1
4 cups	rhubarb	1 L
1 cup	sugar	250 mL
1 tbsp.	flour	15 mL
	pastry for 9-inch (23-cm), double-crust pie	

Preheat oven to 350°F (180°C). Line a 9-inch (23-cm) pie plate with pastry. Beat egg white until stiff. Combine rhubarb, sugar, and flour; fold into egg white. Put filling in pie crust. Cut vents in top pastry layer to allow steam to escape. Place on filled pie crust and flute edges. Bake for 60 minutes.

Rhubarb Parfait Pie

Makes 8 servings.

1	9-inch (23-cm) pie shell, baked and cooled	1
3 cups	rhubarb, diced	750 mL
½ cup	sugar	125 mL
½ cup	water	125 mL
package	strawberry-flavoured gelatin	package
2 cups	vanilla ice cream	500 mL
	coconut (optional)	

Combine rhubarb, sugar, and water in saucepan. Bring to boil; reduce heat to low for 10 minutes. Dissolve gelatin in hot rhubarb. Add ice cream by spoonfuls. Stir until melted. Chill until thickened but not set (10 to 15 minutes). Turn into pie shell. Chill until firm. Garnish with coconut, if desired.

Did you know?

Rhubarb plants will enjoy the addition of well-rotted manure to their soil. It can be applied in either spring or fall.

Rhubarb-Strawberry Pie

Makes 8 servings.

$\frac{3}{4}$ cup	brown sugar	180 mL
$\frac{1}{2}$ cup	sugar	125 mL
2 tbsp.	flour	30 mL
1 tsp.	lemon zest	5 mL
2 cups	strawberries, sliced	500 mL
2 cups	fresh rhubarb, in 1/4-inch (6-mm) pieces	500 mL
	pastry for 9-inch (23-cm), double-crust pie	

Preheat oven to 375°F (190°C). Place one pie crust in pie pan. Combine sugars, flour, and lemon zest. Toss lightly with fruit. Place in bottom pie shell. Cover with top crust. Bake 50 minutes.

Grandma's Rhubarb-Strawberry Pie

Makes 8 servings.

1 $\frac{1}{4}$ cups	sugar	310 mL
1 tbsp.	tapioca	15 mL
1 tbsp.	strawberry-flavoured gelatin	15 mL
4 cups	rhubarb	1 L
2 tbsp.	butter	30 mL
	pastry for 9-inch (23-cm), double-crust pie	

Preheat oven to 450°F (230°C). Place crust in pie pan. Combine sugar, tapioca, and gelatin. Sprinkle 1 tbsp. (15 mL) of this mixture over crust. Combine the rest with the rhubarb in a bowl, stirring occasionally while top crust is being prepared. Spoon rhubarb mixture into pie shell. Dot with butter. Cover with top crust. Bake for 10 minutes, then reduce heat to 350°F (180°C) and bake for 30 minutes more, or until crust is brown and rhubarb is tender.

Rhubarb Lattice Pie

Makes 8 servings.

1 cup	sugar	250 mL
3 tbsp.	flour	45 mL
1 tsp.	orange zest	5 mL
1 tbsp.	butter	15 mL
2	eggs, beaten	2
3 cups	rhubarb, chopped	750 mL
2 tbsp.	butter	30 mL
	pastry for 9-inch (23-cm), double-crust pie	

Preheat oven to 450°F (230°C) and prepare pastry for 2 crusts. Mix together the sugar, flour, orange zest, and butter. Add the eggs and beat until smooth. Spread rhubarb over bottom pie crust. Pour egg mixture evenly over rhubarb. Roll out remaining pastry, sprinkle with sugar, and cut into lattice strips. Arrange over pie; dot with butter. Bake for 10 minutes. Reduce heat to 350°F (180°C) and bake 30 to 40 minutes more.

Ruby Rhubarb Pie

Makes 6 to 8 servings.

3	eggs	3
1/4 cup	sugar	60 mL
1/4 cup	butter	60 mL
3 tbsp.	concentrated orange juice	45 mL
1/4 cup	flour	60 mL
1/4 tsp.	salt	1 mL
1 cup	sugar	250 mL
2 1/2 cups	rhubarb, diced	625 mL
3/4 tsp.	nutmeg	3 mL
	unbaked 9-inch (23-cm) pie shell	

Preheat oven to 375°F (190°C). Separate eggs. Beat egg whites until stiff. Beat in 1/4 cup (60 mL) sugar and set aside. Combine butter, orange juice, flour, egg yolks, salt, and 1 cup (250 mL) sugar. Beat well. Add rhubarb and nutmeg, and fold into beaten egg whites. Put into pie shell and bake pie on bottom rack of oven for 15 minutes. Reduce heat to 325°F (160°C) and continue baking for 35 minutes. Serve warm or at room temperature with ice cream.

Variation: *Use 1 1/2 cups (375 mL) rhubarb and 1 1/2 cups (375 mL) saskatoon berries in place of rhubarb.*

Rhuberry Pie

Makes 8 servings.

$2\frac{1}{2}$ cups	strawberries, halved	625 mL
$2\frac{1}{2}$ cups	fresh rhubarb, chopped	625 mL
1	orange zest	1
$1\frac{1}{2}$ cups	sugar	375 mL
3 tbsp.	flour	45 mL
2 tsp.	tapioca	10 mL
3 tbsp.	butter	45 mL
	pastry for 9-inch (23-cm), double-crust pie	

Preheat oven to 425°F (220°C). Prepare bottom crust for a deep 9-inch (23-cm) pie dish. In a mixing bowl, gently but thoroughly combine strawberries, rhubarb, orange zest, sugar, flour, and tapioca. Pour half of mixture into prepared pie shell. Dot with half of butter. Add rest of fruit and dot with butter. Top with pastry that has been slit to release steam. Bake for 15 minutes, then reduce temperature to 375°F (190°C) and continue baking for another 30 to 40 minutes.

Pie Plant-Peach Pie

1 $8\frac{1}{2}$-oz. can	sliced peaches	265-mL can
2 cups	rhubarb, diced	500 mL
$\frac{1}{4}$ cup	coconut, flaked	60 mL
$1\frac{1}{4}$ cups	sugar	310 mL
3 tbsp.	tapioca	45 mL
1 tsp.	vanilla extract	5 mL
3 tbsp.	butter or margarine	45 mL
	pastry for 9-inch (23-cm), double-crust pie	

Preheat oven to 325°F (160°C). Drain peaches, reserving liquid. Combine with rhubarb, coconut, sugar, and tapioca. Add peach juice and vanilla. Fill one pastry shell with mixture. Dot with butter. Add pastry top and seal edges. Make slits for steam to escape. Bake 45 to 50 minutes until top is golden.

Banana-Rhubarb Pie

Makes 8 servings.

3 cups	rhubarb, diced	750 mL
3	bananas, medium	3
1 cup	sugar	250 mL
$\frac{1}{4}$ cup	orange juice	60 mL
3 tbsp.	flour	45 mL
$\frac{1}{4}$ tsp.	salt	1 mL
$\frac{1}{4}$ tsp.	cinnamon	1 mL
$\frac{1}{4}$ tsp.	nutmeg	1 mL
1 tbsp.	butter	15 mL
	pastry for 9-inch (23-cm), double-crust pie	

Preheat oven to 450°F (230°C). Line 9-inch (23-cm) pie plate with pastry. In large bowl, combine rhubarb, bananas, sugar, orange juice, flour, salt, and spices. Pour into pie shell. Dot with butter. Place remaining pastry over top of filling. Seal, flute, and bake for 15 minutes. Reduce heat to 350°F (180°C) and bake until golden-brown.

Rhubarb-Raisin Pie

Makes 8 servings.

$\frac{1}{3}$ cup	raisins	80 mL
	red wine	
2 cups	rhubarb	500 mL
2 tbsp.	flour	30 mL
6 tbsp.	sugar	90 mL
2 tbsp.	cream	30 mL
1 tbsp.	sugar	15 mL
	shortcrust pastry for 9-inch (23-cm), double-crust pie	

Preheat oven to 425°F (220°C). Place raisins in a small dish and add red wine just to cover. Let soak. Line a pie plate with half the pastry; cut rhubarb into $\frac{1}{2}$-inch (12-mm) pieces and pour into pie shell. Sprinkle rhubarb with flour and 6 tbsp. (90 mL) of sugar. Spread the raisins over top of the rhubarb and pour the red wine over all. Cover with the top pastry and cut vents. Brush with cream and sprinkle 1 tbsp. (15 mL) of sugar over the top. Bake for 15 minutes, then lower temperature to 350°F (180°C) and bake for a further 20 to 25 minutes. Cover loosely with foil if pastry is browning too quickly.

Hawaiian Rhubarb Pie

Makes 8 servings.

2	eggs	2
1 cup	sugar	250 mL
1/4 tsp.	salt	1 mL
2 1/2 tbsp.	flour	37 mL
1/2 tsp.	vanilla	2 mL
2 tbsp.	milk	30 mL
2 cups	rhubarb, diced	500 mL
1 cup	pineapple, crushed, drained	250 mL
1 tbsp.	strawberry-flavoured gelatin	15 mL
	pastry for a 9-inch (23-cm), double-crust pie	

Preheat oven to 350°F (180°C). Place crust in pie pan. Beat eggs. Add sugar, salt, flour, vanilla, and milk. Beat well. Pour over rhubarb and crushed pineapple, stirring lightly. Spoon rhubarb mixture into pie shell. Cover with top crust. Bake for 35 minutes or until crust is brown and rhubarb is tender.

Strawberry-Rhubarb Mousse Pie

Makes 10 servings.

3 1/2 cups	rhubarb, diced	875 mL
1 cup	sugar	250 mL
1/4 cup	water	60 mL
2 packages	unflavoured gelatin	2 packages
1/2 cup	cold water	125 mL
2 cups	strawberries	500 mL
1 tbsp.	lemon juice	15 mL
6 tbsp.	butter or margarine, melted	90 mL
2 cups	shortbread cookie crumbs	500 mL
1 cup	whipping cream	250 mL
	mint sprigs	
	strawberries	

Preheat oven to 350°F (180°C). In 2-quart (2-L) saucepan, heat rhubarb, sugar, and 1/4 cup (60 mL) water to boiling over high heat, stirring constantly. Reduce heat to medium-low and continue cooking until rhubarb is very tender, about 10 minutes.

In food processor, process rhubarb until smooth; return to pan. • Sprinkle gelatin over ½ cup (125 mL) water; let stand 2 minutes to soften. In another bowl, smash strawberries with potato masher or fork. Stir softened gelatin, mashed strawberries, and lemon juice into rhubarb; cook 3 minutes. Pour rhubarb mixture into bowl and refrigerate, stirring occasionally, until mixture mounds slightly when dropped from a spoon, about 2½ hours (or, for quicker setting, place in a larger bowl of ice water and stir every 10 minutes until mixture mounds, about 1 hour). • Place melted butter in a deep-dish 9-inch (23-cm) pie pan and mix with cookie crumbs. Press mixture onto bottom and up sides. Bake 15 minutes until golden; cool. Whip cream until stiff peaks form. Fold into chilled rhubarb mixture. Spoon into pie shell. Refrigerate 3 hours or overnight, until pie is well chilled. Garnish with mint and strawberries to serve.

Strawberry-Rhubarb Meringue Pie

Makes 8 servings.

3 tbsp.	quick-cooking tapioca	45 mL
2 cups	strawberries, sliced	500 mL
5 cups	rhubarb, diced	1.25 L
¾ cup	sugar	180 mL
2 tbsp.	water	30 mL
5	egg whites, large	5
¼ tsp.	cream of tartar	1 mL
⅓ cup	granulated sugar, fine	80 mL
1	9-inch (23-cm) graham-cracker crust, baked	1

Preheat oven to 350°F (180°C). Stir tapioca, strawberries, rhubarb, and ¾ cup (180 mL) sugar together with 2 tbsp. (30 mL) water in a non-reactive saucepan. Let sit for 5 minutes. Cover and cook mixture over low heat, stirring occasionally, until rhubarb begins to release liquid, about 10 minutes. Cook on medium-low heat until mixture begins to bubble gently. Then uncover and simmer, stirring constantly, until rhubarb softens and mixture thickens, about 10 minutes longer. Set aside to cool slightly while you prepare the meringue topping. • Beat egg whites and cream of tartar with an electric mixer in a large, clean bowl until soft peaks form. Slowly beat in ⅓ cup (80 mL) sugar. • Pour the warm rhubarb filling into the prepared crust. Spread the meringue over the filling, mounding it toward the center. The meringue should cover the edges of the crust. Bake in the pre-heated oven for about 15 minutes, until top is evenly golden. Cool for one hour. Refrigerate and serve cold.

Rhubarb-Orange Cream Pie

Makes 8 servings.

3	eggs, separated	3
1/4 cup	sugar	60 mL
1/4 cup	butter, softened	60 mL
3 tbsp.	orange juice, frozen concentrate	45 mL
1 cup	sugar	250 mL
1/4 cup	flour	60 mL
1/4 tsp.	salt	1 mL
2 1/2 cups	rhubarb, chopped	625 mL
1/3 cup	pecans, chopped	80 mL
	pastry for 9-inch (23-cm), single-crust pie	

Preheat oven to 350°F (180°C). Prepare pastry shell; set aside. Beat egg whites until stiff but not dry. Add 1/4 cup (60 mL) sugar gradually, beating until whites are stiff and glossy; set aside. • Beat butter, thawed orange juice concentrate, and egg yolks until well blended. Add remaining sugar, flour, and salt. Stir in rhubarb, then gently fold in egg whites. Pour into prepared shell; sprinkle top with nuts. Place pie on bottom rack of oven and bake for 15 minutes. Reduce temperature to 325°F (160°C) and bake for a further 45 minutes.

Rhubarb Cream Pie

Makes 8 servings.

1 1/4 cups	sugar	310 mL
3 tbsp.	flour	45 mL
1/2 tsp.	nutmeg	2 mL
1 tbsp.	butter	15 mL
2	eggs, well beaten	2
3 cups	rhubarb, chopped	750 mL
	pastry for 9-inch (23-cm), double-crust pie	

Preheat oven to 450°F (230°C). Blend sugar, flour, nutmeg, and butter. Add eggs and beat until smooth; pour over rhubarb in pastry-lined pan. Top with pastry; cut pastry to release steam, as desired. Bake for 10 minutes, then reduce heat and bake at 350°F (180°C) for about 30 minutes.

Creamy Rhubarb Pie

Makes 8 to 9 servings.

2 cups	rhubarb	500 mL
3	egg yolks	3
½ cup	half-and-half cream **or** milk	125 mL
⅔ cup	sugar	160 mL
2 tbsp.	flour	30 mL
3	egg whites	3
¼ tsp.	cream of tartar	1 mL
⅓ cup	sugar	80 mL
½ tsp.	vanilla	2 mL
	pastry for 9-inch (23-cm), single-crust pie	

Preheat oven to 375°F (190°C). On a lightly floured board, roll pastry into a 12-inch (30-cm) diameter round; ease into a 9-inch (23-cm) pie pan. Fold edges under; flute rim decoratively. • Cut rhubarb into ½-inch (12-mm) pieces and place in pastry. In a bowl, beat yolks, cream, ⅔ cup (160 mL) sugar, and flour until smooth; pour over rhubarb. Bake on lowest rack in oven until pastry is golden-brown and custard appears set in center when pie is jiggled, 40 to 45 minutes. If rim begins to darken excessively, drape with foil. • In a large bowl, beat egg whites and cream of tartar at high speed with an electric mixer until frothy. Gradually whip in ⅓ cup (80 mL) sugar and vanilla, beating until whites hold stiff, glossy peaks. Pile meringue onto hot filling. With a spatula, swirl meringue over filling and up against rim of pastry. Bake at 400°F (200°C) until meringue is tinged with brown, 3 to 5 minutes. (If you are concerned about undercooked eggs, continue to bake until centre of meringue is 160°F (70°C), about 7 minutes.) Serve warm or at room temperature.

Hint: *If making ahead, let pie cool, then cover without touching meringue and chill for up to a day. Cut into wedges.*

Rhubarb Custard Pie

Makes 8 servings.

3 cups	rhubarb, diced	750 mL
1 ½ cups	sugar	375 mL
3 tbsp.	flour	45 mL
pinch	salt	pinch
½ tsp.	nutmeg	2 mL
2	eggs	2
2 tbsp.	milk	30 mL
1 tbsp.	butter	15 mL
	pastry for 9-inch (23-cm), single-crust pie	

Preheat oven to 400°F (200°C). Mix rhubarb with sugar, flour, salt, and nutmeg. Beat eggs slightly; add milk. Combine with rhubarb mixture. Place in pie shell and dot with butter. Bake for 50 to 60 minutes.

Anna's Rhubarb Custard Pie

Makes 8 servings.

3 cups	rhubarb, diced	750 mL
2 tbsp.	flour	30 mL
1 ½ cups	sugar	375 mL
1 tbsp.	butter, melted	15 mL
3	eggs, separated	3
	cinnamon	
	pastry for 9-inch (23-cm), single-crust pie	

Preheat oven to 450°F (230°C). Place rhubarb in a mixing bowl. Combine flour and sugar, and stir into rhubarb with melted butter. Beat egg yolks and add to rhubarb mixture. Beat egg whites until stiff but not dry; fold into rhubarb mixture. Spoon into unbaked pie shell and sprinkle with cinnamon. Bake for 10 minutes. Reduce heat to 350°F (180°C) and bake for another 30 to 40 minutes, until golden-brown.

Quick Rhubarb Custard Pie

Makes 8 to 10 servings.

3	eggs	3
8 tsp.	milk	40 mL
1 to 2 cups	sugar	250 to 500 mL
4 tbsp.	flour	60 mL
¾ tsp.	nutmeg	4 mL
4 cups	rhubarb	1 L
	butter	
	pastry for 9-inch (23-cm), single-crust pie	

Preheat oven to 425°F (220°C). Mix eggs, milk, sugar, flour, and nutmeg. Cut rhubarb in ½-inch (12-mm) pieces and stir into egg mixture; pour into pie shell. Dot with butter. Bake for 15 minutes. Lower oven temperature to 375°F (190°C) and bake for a further 45 minutes or until crust is golden-brown and rhubarb is tender.

Rhubarb Custard Pie with Crumble Topping

Makes 8 servings.

3½ cups	rhubarb, diced	875 mL
1½ cups	sugar	375 mL
3	eggs, beaten	3
½ cup	cream **or** half-and-half	125 mL
½ tsp.	salt	2 mL
1 tbsp.	flour	15 mL
1½ cups	quick-cooking oats	375 mL
1 cup	brown sugar	250 mL
½ tsp.	nutmeg	2 mL
¼ cup	butter, softened	60 mL
	pastry for 9-inch (23-cm), single-crust pie	

Preheat oven to 400°F (200°C). Combine rhubarb, sugar, eggs, cream, and salt. Sprinkle flour over bottom of pie shell. Pour filling into shell. • Combine oats, brown sugar, nutmeg, and butter; sprinkle over filling. Bake for 10 minutes. Reduce heat to 350°F (180°C) and continue to bake for a further 50 minutes.

Orange Custard and Rhubarb Ripple Pie

Makes 8 servings.

1	9-inch (23-cm) graham-cracker crust, baked	1
1½ cups	whole milk	375 mL
2	eggs	2
1	egg yolk	1
½ cup	sugar	125 mL
3 tbsp.	flour	45 mL
1 tsp.	vanilla	5 mL
1½ tsp.	orange zest	7 mL
1 tsp.	Grand Marnier	5 mL
3 cups	rhubarb, diced	750 mL
3 tbsp.	sugar	45 mL
3 tbsp.	orange juice	45 mL
1 cup	heavy whipping cream	250 mL
1 tsp.	vanilla	5 mL
1 tbsp.	icing sugar	15 mL

Heat the milk in a heavy medium-size saucepan until the milk is hot and a few bubbles form around the edge; ***do not let it boil****. Whisk the eggs, egg yolk, ½ cup (125 mL) sugar, and flour until smooth. Slowly whisk the hot milk into the egg mixture. Return the mixture to the saucepan and cook over medium heat, stirring constantly, until the mixture boils and thickens. Transfer to a bowl. Stir in the vanilla, orange zest, and Grand Marnier. Press plastic wrap on the surface to keep from drying and refrigerate the mixture until cold. (The filling can be prepared up to one day ahead and kept refrigerated.) • Cook the rhubarb, 3 tbsp. (45 mL) sugar, and orange juice in a covered non-reactive saucepan over low heat until it begins to release liquid, about 5 minutes. Uncover and simmer over medium heat, stirring constantly, until the rhubarb softens, about 5 minutes longer. Set aside to cool slightly for about 10 minutes, then process the mixture in a food processor until smooth. Cover and refrigerate until cold. • Reserve 1/4 cup (60 mL) of the rhubarb sauce. Spread the remaining sauce over the prepared crust. Spread the cold orange custard over the rhubarb sauce. Drizzle the reserved rhubarb sauce over the custard, swirling it gently into the custard. Whip cream with vanilla and icing sugar until firm peaks form. Spread the whipped cream over the top of the pie. Serve cold.*

Hint: *This pie is best served the day it is prepared.*

Rhubarb Meringue Custard Pie

Makes 6 servings.

3 cups	rhubarb	750 mL
1 cup	sugar	250 mL
3 tbsp.	flour	45 mL
2 tbsp.	butter, cubed	30 mL
2	egg yolks	2
2	egg whites	2
1/4 tsp.	cream of tartar	1 mL
1/4 cup	sugar	60 mL
2 tbsp.	water	30 mL
1/2 tsp.	vanilla	2 mL
1/4 tsp.	salt	1 mL
	pastry for 9-inch (23-cm), single-crust pie	

Preheat oven to 425°F (220°C). Cut rhubarb into 1-inch (2.5-cm) pieces and combine with 1 cup (250 mL) sugar, flour, and butter in large bowl. In separate bowl, beat egg yolks until light; stir into rhubarb mixture. Spoon rhubarb mixture into pie shell. Bake in oven 10 minutes. Reduce temperature to 350°F (180°C); bake 30 minutes or until filling is bubbling and rhubarb is tender. Remove from oven and let cool to lukewarm on wire rack. • Increase oven temperature to 375°F (190°C). In a large bowl, use an electric mixer to beat egg whites with cream of tartar 3 to 5 minutes, until stiff, moist peaks form. Very gradually beat in 1/4 cup (60 mL) sugar. Add water, vanilla, and salt; beat until very stiff, shiny peaks form. • Spread meringue over lukewarm pie, making sure it touches pastry edges all the way around. Swirl meringue decoratively with knife. Bake 10 to 12 minutes until tips of meringue are golden-brown. Let pie cool slowly on wire rack. Serve at room temperature.

Did you know?

Sandy soils are not recommended for growing rhubarb. The plant thrives in moist soil conditions, but needs a well-drained soil.

Rhubarb Meringue Pie

Makes 8 servings.

4 cups	rhubarb, finely chopped	1 L
3	egg yolks	3
1 cup	sugar	250 mL
2 tbsp.	flour	30 mL
2 tbsp.	melted butter	30 mL
3	egg whites	3
⅓ cup	sugar	80 mL
	pastry for 9-inch (23-cm), single-crust pie	

Preheat oven to 350°F (180°C). Combine rhubarb and egg yolks. Add 1 cup (250 mL) sugar and flour and mix well. Stir in butter and spoon into pie shell. Bake for 30 minutes or until set. • Beat egg whites until soft peaks form. Continue beating and slowly add ⅓ cup (80 mL) sugar until mixture is thick and glossy. Heap lightly over pie and spread to edges of the crust. Turn oven up to 400°F (200°C) and bake pie for about 10 to 12 minutes, until peaks of meringue turn golden-brown.

Sour Cream-Rhubarb Pie

Makes 8 servings.

2 cups	rhubarb, diced	500 mL
2 cups	sugar	500 mL
1 tbsp.	flour	15 mL
1 cup	sour cream	250 mL
¼ tsp.	vanilla extract	1 mL
pinch	salt	pinch
½ cup	brown sugar	125 mL
¼ cup	butter	60 mL
⅓ cup	flour	80 mL
	pastry for 9-inch (23-cm), single-crust pie	

Preheat oven to 400°F (200°C). Mix rhubarb, sugar, flour, sour cream, vanilla, and salt together and place in unbaked pie shell. Bake 20 to 25 minutes; remove from oven. • Mix brown sugar, butter, and flour together and spread on top of pie. Bake for a further 20 to 25 minutes.

Rhubarb Custard Torte

Makes 8 servings.

¾ cup	butter	180 mL
⅓ cup	sugar	80 mL
2	egg yolks	2
2 cups	flour	500 mL
1 tsp.	baking powder	5 mL
½ tsp.	salt	2 mL
6 cups	rhubarb	1.5 L
½ cup	sugar	125 mL
¼ cup	quick-cooking tapioca	60 mL
½ tsp.	cinnamon	2 mL
¼ cup	water	60 mL
6	eggs	6
2 cups	sour cream	500 mL
½ cup	brown sugar, packed	125 mL
2 tsp.	lemon zest	10 mL
1 tsp.	vanilla	5 mL
	icing sugar	
	lemon zest	

Preheat oven to 400°F (200°C). In large bowl, cream butter and ⅓ cup (80 mL) sugar; add egg yolks and beat until light and fluffy. In separate bowl, sift together flour, baking powder, and salt; add to egg mixture, mixing until crumbly. Press ⅔ of mixture into bottom of 10-inch (3-L) springform pan. Bake for 10 minutes or just until lightly golden; cool. Press remaining mixture up sides of pan. • Cut rhubarb into 1-inch (2.5-cm) pieces. In heavy stainless-steel saucepan, stir together rhubarb, ½ cup (125 mL) sugar, tapioca, and cinnamon. Let stand for 15 minutes. Stir in water and bring to boil. Reduce heat to medium-low; cook, covered and stirring often, for about 10 minutes or just until rhubarb is tender but not mushy. (Mixture should be quite thick.) Let cool slightly; pour over crust. • In large bowl, whisk eggs for about 2 minutes or until frothy; stir in sour cream, brown sugar, lemon zest, and vanilla. Pour over rhubarb and bake in 350°F (180°C) oven for about 1 hour or until top is golden and custard has set. Let cool and refrigerate, covered, for at least 3 hours or up to 12 hours. Run sharp knife between pan and crust before removing outside ring of pan. Sprinkle icing sugar and lemon zest on center of the torte.

Brown Sugar-Rhubarb Pie

Makes 8 servings.

1	9-inch (23-cm) pie shell, unbaked	1
1 1/4 cups	brown sugar, firmly packed	310 mL
1/4 cup	flour	60 mL
1/4 tsp.	salt	1 mL
4 cups	rhubarb, diced	1 L
3	egg yolks	3
1 tbsp.	lemon juice	15 mL
3	egg whites	3
1/2 tsp.	vanilla	2 mL
1/4 tsp.	cream of tarter	1 mL
6 tbsp.	sugar	90 mL

Preheat oven to 450°F (230°C). Bake pastry for 5 minutes and cool. • Combine brown sugar, flour, and salt. Add to rhubarb; toss to coat fruit. Let stand for 15 minutes. Beat egg yolks slightly with fork. Stir yolks and lemon juice into rhubarb mixture. Turn rhubarb filling into the partially baked pastry shell. Cover edge of pie with foil. Reduce oven to 375° (190°C) and bake for 25 minutes. Remove foil; bake for 20 to 25 minutes more or till nearly set. (Pie becomes firm after cooling.) • Towards end of baking time, prepare meringue by placing egg whites, vanilla, and cream of tarter in small mixing bowl and beating with electric mixture on medium speed until soft peaks form. Gradually add the sugar, a spoonful at a time. Beat at high speed for about 4 minutes or until the mixture forms stiff peaks and the sugar is dissolved. • Spread meringue over hot rhubarb filling, making sure to seal to edge of pastry. Reduce oven to 350°F (180°C) and bake for 12 to 15 minutes. Cool. Cover and chill to store.

Rhubarb-Strawberry Tart with Orange Shortbread Crust

Makes 8 or 9 servings.

1 ⅓ cups	flour	330 mL
3 tbsp.	sugar	45 mL
½ cup	butter or margarine	125 mL
1 ½ tsp.	orange zest	7 mL
1	egg yolk	1
4 cups	strawberries, halved	1 L
3 cups	rhubarb, diced	750 mL
⅔ cup	sugar	160 mL
1 tsp.	orange zest	5 mL
¼ cup	cornstarch	60 mL
2 tbsp.	water	30 mL
2 tbsp.	orange liqueur (optional)	30 mL

Preheat oven to 325°F (160°C). In a food processor or by hand, cut together flour, sugar, butter, and 1½ tsp. (7 mL) orange zest. Add egg yolk and process until dough holds together. Press evenly over bottom and up sides of an 11-inch (28-cm) springform pan. Bake until golden, 25 to 30 minutes; cool. • Set aside 1 cup (250 mL) of strawberries for garnish. In a 2 to 3-quart (2 to 3-L) pan over medium heat, combine remaining strawberries, rhubarb, sugar, and 1 tsp. (5 mL) orange zest. Cover and stir occasionally until rhubarb is soft when pierced, about 5 minutes. Blend water with cornstarch; stir into rhubarb mixture. On high heat, stir until mixture reaches a rolling boil. Set aside. Add liqueur to rhubarb mixture, then spread into baked tart shell. Chill until filling is cool and set, at least 1 hour; if making ahead, cover and chill for up to a day. • Top with reserved strawberries. Cut into wedges and serve.

Strawberry-Rhubarb Meringue Tart

Makes 8 servings.

1½ cups	flour	375 mL
3 tbsp.	sugar	45 mL
¼ tsp.	salt	1 mL
¼ cup	unsalted butter	60 mL
¼ cup	vegetable shortening	60 mL
3 tbsp.	fresh orange juice	45 mL
1 tbsp.	orange liqueur	15 mL
2 tbsp.	sugar	30 mL
4 cups	rhubarb	1 L
2 cups	strawberries, halved	500 mL
1½ cups	sugar	375 mL
¼ cup	quick-cooking tapioca	60 mL
4 tsp.	orange zest	20 mL
¼ tsp.	salt	1 mL
2 tbsp.	unsalted butter	30 mL
1 tbsp.	orange liqueur	15 mL
4	egg whites	4
¼ tsp.	cream of tartar	1 mL
½ cup	sugar	125 mL
2 tbsp.	sugar	30 mL

Combine flour, 3 tbsp. (45 mL) sugar, and ¼ tsp. (1 mL) salt in large bowl. Cut in butter and shortening until coarse meal forms. Make well in center of dry ingredients. Add orange juice and liqueur. Toss with fork to combine. Knead dough just until it holds together. Form into ball; flatten. Wrap with plastic and refrigerate 30 minutes (or prepare a day ahead). • Preheat oven to 400°F (200°C). Grease an 11-inch (28-cm) springform pan. Roll dough out on lightly floured surface to 14-inch (35-cm) round. Transfer to prepared pan. Trim and finish edges. Pierce bottom of crust with fork. Chill for 30 minutes. Line crust with foil and fill with rice. Bake on lower rack of oven until dough is set, about 15 minutes. Remove foil and rice. Sprinkle crust with sugar. Bake until brown, about 15 minutes. Cool on rack. • Cut rhubarb into ½-inch (12-mm) pieces and combine thoroughly with strawberries, 1½ cups (375 mL) sugar, tapioca, orange zest, and ¼ tsp. (1 mL) salt in large, heavy saucepan. Let stand 30 minutes, mixing occasionally. Bring to boil, stirring constantly. Reduce heat to medium and cook until liquid thickens and rhubarb softens, stirring occasionally. Remove from heat. Mix in butter and liqueur. Cool completely. (Can prepare day before and refrigerate.) • Preheat oven to 425°F (220°C). Using electric mixer, beat whites until foamy. Add cream of tartar and beat until soft peaks form. Add ½ cup (125 mL) sugar a spoonful at a time, beating until stiff and glossy. • Turn fruit filling into crust. Pipe meringue decoratively on top of tart using a pastry bag, covering completely and sealing to edge all around. Sprinkle with 2 tbsp. (30 mL) sugar. Bake until meringue is light brown, about 5 minutes. Cool completely. Serve at room temperature.

Rhubarb Cheesecake Pie

Makes 9 to 12 servings.

18	graham crackers	18
3 tbsp.	butter or margarine	45 mL
¾ cup	sugar	180 mL
3 tbsp.	cornstarch	45 mL
4 cups	rhubarb	1 L
1 tbsp.	water	15 mL
6 oz.	cream cheese	170 g
2	eggs	2
½ tsp.	vanilla	2 mL
6 tbsp.	sugar	90 mL
1 cup	sour cream	250 mL
1 tbsp.	sugar	15 mL

Preheat oven to 350°F (180°C). In food processor, whirl graham crackers to make 1 cup (250 mL) fine crumbs. Pour crumbs into a 9-inch (23-cm) pie pan. Melt butter and mix with crumbs. Press mixture firmly over bottom and sides of pan. Bake until darker brown at rim, 8 to 10 minutes. • In a 2 to 3-quart (2 to 3-L) pan, mix ¾ cup (180 mL) sugar and cornstarch. Cut rhubarb into 1-inch (2.5-cm) pieces and add to cornstarch mixture with water. Stir often over medium heat until mixture comes to a full boil. Pour rhubarb mixture into crust. With a mixer, blend cream cheese, eggs, vanilla, and 6 tbsp. (90 mL) sugar until smooth; pour over rhubarb. Bake until filling appears set in centre when pan is gently shaken, about 20 minutes. • Mix sour cream with 1 tbsp. (15 mL) sugar; spread evenly over filling. Bake until topping is set when gently shaken, about 5 minutes. Let cool, cover, and chill at least 2 hours or until next day. Cut into wedges and serve.

Rhubarb Crumb Tart

Makes 8 or 9 servings.

1	large egg, cold	1
1 tsp.	vanilla	5 mL
1 cup	flour	250 mL
¼ cup	cake flour	60 mL
¼ cup	sugar	60 mL
⅓ cup	confectioner's sugar	80 mL
½ tsp.	baking powder	2 mL
⅛ tsp.	salt	0.5 mL
¼ lb.	unsalted butter, cut in pieces	114 g
1½ cups	flour	375 mL
1½ cups	brown sugar, packed	375 mL
¼ tsp.	salt	1 mL
1 tsp.	cinnamon	5 mL
7 tbsp.	unsalted butter, melted	105 mL
1 tsp.	vanilla extract	5 mL
3 cups	rhubarb, diced	750 mL
¼ cup	sugar	60 mL
3 tbsp.	water	45 mL

Grease an 11-inch (28-cm) metal springform pan. Mix the egg and vanilla in a small bowl; set aside. Mix both flours, both sugars, baking powder, and salt in the large bowl of an electric mixer on low speed just to blend them. Mix in the butter pieces until they are the size of small lima beans, about 30 seconds. With the mixer running, add the egg mixture. Stop the mixer as soon as the mixture begins to hold together. Form the dough into a smooth round disk. Wrap in plastic wrap. Refrigerate for 1 hour or overnight. • Stir 1½ cups (375 mL) of flour, brown sugar, salt, and cinnamon together in a large bowl. Mix the melted butter and vanilla together in a small bowl, then stir it into the flour mixture until crumbs form. Set aside. • Stir the rhubarb and ¼ cup (60 mL) sugar together with water in a non-reactive saucepan. Cover and cook the mixture over low heat until it comes to a simmer and releases liquid, about 5 minutes. Remove the cover and boil gently, stirring often, until the rhubarb softens and the mixture thickens, about 5 minutes. Set aside to cool while you roll out the chilled dough. • Roll the chilled dough out on a lightly floured surface to a 14-inch (35-cm) circle. Transfer to the tart pan. Trim the edges to leave a ¾-inch (18-mm) overhang. Fold the overhang into the pan and press it against the edge of the pan. Refrigerate until cold, about 1 hour. • To bake, preheat oven to 375°C (190°F). Press a piece of heavy aluminum foil into the cold crust. Fill the foil with raw rice, dried beans, or metal pie weights. Bake the crust for 20 minutes. Carefully remove the foil and weights. Reduce the oven temperature to 350°C (180°F). Spread the rhubarb sauce evenly over the crust. Sprinkle the reserved crumb topping over the filling. Bake for about 25 minutes until the crust is golden. Cool and serve at room temperature.

***Hint:** The tart can be prepared up to 3 days in advance.*

Rhubarb Tart

Makes about 8 servings.

9 cups	rhubarb, finely diced	2.25 L
1¼ cups	sugar	310 mL
2 tbsp.	orange zest	30 mL
1¼ cups	flour	310 mL
1 tbsp.	sugar	15 mL
¼ tsp.	salt	1 mL
1 tbsp.	butter	15 mL
¼ cup	vegetable oil	60 mL
2 to 3 tbsp.	ice water	30 to 45 mL

Preheat oven to 400°F (200°C). Set a large baking sheet on rack in lower third of oven to preheat. Lightly oil a 10½-inch (26-cm) springform pan or coat it with non-stick cooking spray; set aside. • Cover and set aside 3 cups (750 mL) of the rhubarb. In a heavy saucepan, combine the remaining rhubarb, 1¼ cups (310 mL) sugar, and orange zest. Bring to a boil, reduce heat, and simmer gently for about 40 to 50 minutes, until the purée has reduced to about 1⅔ cups (410 mL). Stir often to prevent scorching. Let cool to room temperature. • Stir together flour, 1 tbsp. (15 mL) sugar, and salt. In a small saucepan, melt butter over low heat. Cook, swirling the pan, until butter turns a nutty brown, about 30 seconds. Pour the butter into a small bowl and stir in oil. Using a fork, slowly stir the butter-oil mixture into the dry ingredients until crumbly. Gradually stir in ice water until the dough holds together and is not at all crumbly. Roll out the dough into a circle about 14 inches (35 cm) in diameter. Invert the dough into the springform pan. Fold in the overhanging edge of the crust to form a sturdy edge, patching any thin spots with scraps. • Combine the reserved raw and cooked rhubarb and spread in the crust. Place the tart pan on the preheated cookie sheet and bake for 35 to 40 minutes, or until the crust is golden and the filling bubbles. Let the tart cool to room temperature before serving. This pie is great topped with vanilla frozen yogurt and a dusting of ground cloves.

***Tip:** Use the reddest rhubarb to give the tart a deep rose colour.*

Ruth's Rhubarb Tart

Makes 8 servings.

1⅔ cups	flour	410 mL
¼ cup	sugar	60 mL
¼ tsp.	salt	1 mL
½ cup	chilled unsalted butter	125 mL
2	egg yolks	2
2 tbsp.	ice water	30 mL
3 tbsp.	apricot jam	45 mL
1 cup	sugar	250 mL
⅓ cup	water	80 mL
3	½-inch (12-mm) strips lemon peel	3
1	cinnamon stick	1
6½ cups	fresh rhubarb (about 2 lbs. or 900 g trimmed)	1.7 L

Preheat oven to 350°F (180°C). Place flour, ¼ cup (60 mL) sugar, and salt in food processor and process until combined. Add butter; pulse until mixture resembles coarse meal. Lightly whisk together egg yolks and ice water; add to flour mixture and process just until moist clumps form. If dough is dry, add more water by spoonfuls to moisten. Shape dough into a flattened disc and refrigerate for 30 minutes or until dough is firm enough to roll. • Roll out dough on floured surface to 12-inch (30-cm) circle. Transfer to 9-inch (23-cm) tart pan with removable bottom. Trim crust overhang to ¼ inch (6 mm). Fold overhang in and press lightly. Put in freezer 15 minutes. Line crust with foil. Fill with dried beans or pie weights. Bake until sides are set, about 20 minutes. Remove foil and weights. Bake for about 15 minutes more or until crust is golden-brown, piercing with fork if bubbles form. Brush crust with jam and bake for an additional 5 minutes or until jam is set. Cool in pan on rack. • Combine 1 cup (250 mL) sugar and water in heavy, 11-inch (28-cm) frypan over low heat. Stir until sugar dissolves. Add lemon zest and cinnamon stick. Bring to boil. Add rhubarb and reduce heat; cover pan and simmer for 5 minutes or until rhubarb is just beginning to soften, stirring occasionally. Remove from heat and let stand for about 15 minutes or until rhubarb is tender, stirring occasionally. Uncover and cool. With slotted spoon, remove rhubarb and drain off as much liquid as possible; set aside. Arrange rhubarb in concentric circles in crust. Strain reserved liquid into small, heavy saucepan. Bring to a boil and cook for 5 to 10 minutes or until reduced to ⅓ cup (80 mL), stirring occasionally. Let cool slightly. Spoon over rhubarb and serve.

cakes, squares, muffins, & breads

more than just pies

Rhubarb's History

Rheum officinale and R. palmatum were described in a Chinese herbal of 2700 BC. At that time, they were used for their strong laxative properties. Many of the species were important in commerce from the Far and Near East and gradually came into European cultivation.

Over the centuries, rhubarb has attracted botanists, horticulturists, business people, doctors, pharmacists, and explorers alike. The ancient Greeks correctly believed that the source of their rhubarb was somewhere in the east. Probably the first European to see and describe R. officinalis was Marco Polo in the latter part of the fifteenth century. He talked about it at length in accounts of his travels in China.

When in Aleppo in 1547, physician Pierre Belon noted that rhubarb root was brought from Mesopotamia at the rate of twelve camel-loads at a time, but he could not find out from what sort of plant this root came. Again at Aleppo in 1547, German physician Leonhardt Rauwolf discovered a species of rhubarb, namely R. ribes (the true Ribes of the Arabians), whose stalks were full of "a pleasant sourish juice." However, it was not the only rhubarb sold by the Arabs. They also sold roots that came from somewhere within the Chinese Empire.

It was not until the seventeenth century that informed Europeans knew that the Chinese Empire was the principal, if not exclusive, source of this tasty treat, although rhubarb is said to have been planted in Italy ca. 1608. Much rhubarb arrived with Russian caravans before 1755. In 1778 rhubarb was reported as a food plant in Europe. During the last quarter of the eighteenth century, the English travelled into the Himalayas from India. Here they found several species of rhubarb, but none proved to be the real medicinal species. By the beginning of the nineteenth century, general intelligence held that the "True" and "Official Rhubarb" came from a district of western

China around Sining. When China was explored in the second half of the nineteenth century, this intelligence proved correct.

Around 1777, an apothecary of Banbury, Oxfordshire named Haywood began growing "Rheum rhaponticum" *from seeds that came from Russia. He produced a drug of excellent quality. This was sold as "genuine rhubarb" by men dressing up as Turks.*

Around 1901, Ernest Henry Wilson explored Omei Shan and Wa Shan in China. He spent much time collecting above and around treeline, and there discovered R. alexandrae. *This is unlike a typical rhubarb in that it is reminiscent of a bolted lettuce. The leaves are large and crinkly, while the flower bracts are a creamy-yellow, rising in tiers to a height between 30 and 200 centimetres. These bracts protect the flowers and, subsequently, the young fruits from the ravages of inclement weather. The bracts quickly wither when the fruits mature, enabling the fruits to be dispersed by the wind. In the early part of the twentieth century, George Foresst also collected this species on three occasions. He found it growing on moist, open, limy pastures and windswept ridges in northwest Yunnan and Tibet at altitudes between 3300 and 4700 metres above sea level. He also collected other rhubarb species including* R. officinalis, R. acuminatum, *and* R. forrestii.

In trade during the last half of the eighteenth and first half of the nineteenth centuries, the Russian State Monopoly and the British East India Company were in fierce competition. Interestingly, this struggle pitted a Russian product of high quality and high price against large amounts of a lower-quality product at a lower price. The Russian rhubarb monopoly ran into problems because of the limited ability of Russian bureaucrats to accommodate the market and to adapt to the realities of western European demand. The Russians also continued to cling to fixed and long-term agreements. These factors led to accumulations of huge stocks that deteriorated over time. The ultimate lesson: there was room in the trade for both a high-quality, high-priced product and one of lower quality and lower price, since after the demise of the monopolies, trade was continued very successfully by private merchants until the opening of the interior of China to European trade.

Rhubarb Shortcake

Makes 6 to 8 servings.

4 cups	rhubarb, diced	1 L
1¼ cups	sugar	310 mL
1 tbsp.	butter (optional)	15 mL
8-inch	prepared shortcake	19-cm
	nutmeg	

Cook rhubarb and sugar until rhubarb is tender. Add butter to hot sauce if desired. Cool. Split the shortcake into 2 layers. Spread half the sauce on bottom layer; add the top layer. Spread the remainder of the sauce on the top. Sprinkle with nutmeg; serve with whipped cream.

Simple Rhubarb Upside-Down Cake

Makes 8 to 10 servings.

5 tsp.	butter or margarine	25 mL
⅔ cup	brown sugar	160 mL
4 cups	rhubarb, chopped	1 L
1 package	white cake mix	1 package

Preheat oven to 350°F (180°C). Prepare a 10 x 6 x 2-inch (25 x 15 x 5-cm) baking dish. Melt butter and stir in sugar. Spread over the bottom of baking dish. Arrange rhubarb on top. Prepare cake mix according to directions. Pour batter over rhubarb to within ½ inch (12 mm) of top of dish. (Use any remaining batter for cupcakes.) Bake for 45 to 50 minutes or until cake is done. Remove from oven and let stand 5 minutes. Invert serving dish over top of cake pan and turn out cake. Serve with whipped cream or ice cream.

Sweet Rhubarb Upside-Down Cake

Makes 8 to 10 servings.

2 to 3 cups	rhubarb, cut up	500 to 750 mL
1 cup	sugar	250 mL
5 cups	miniature marshmallows	250 g
1 package	yellow or white cake mix	1 package

Preheat oven to 350°F (180°C). Grease a 9 x 12-inch (22.5 x 30-cm) pan. Layer the rhubarb into the prepared pan and sprinkle with sugar. Layer the marshmallows over the sugar. Prepare the cake mix according to instructions on the box and spread batter over the marshmallows. Bake 45 minutes or until done. Invert onto serving dish. Cut into squares to serve.

***Option:** 1/2 cup (125 mL) of nuts can be added just before the marshmallow layer.*

Rhubarb Pie Cake

Makes 10 to 12 servings.

3 cups	rhubarb	750 mL
1 cup	water	250 mL
1 cup	sugar	250 mL
5 cups	miniature marshmallows	250 g
1 package	yellow cake mix	1 package
3-oz. package	strawberry-flavoured gelatin	85-g package

Preheat oven to 350°F (180°C). Grease 9 x 13-inch (22.5 x 32-cm) pan. Chop rhubarb into 1-inch (2.5-cm) pieces and mix with water and sugar in saucepan. Bring to boil and set aside. Line pan with miniature marshmallows. Prepare cake mix as directed on package. Pour over marshmallows. Add gelatin to rhubarb mixture and pour over batter. Bake for 40 to 45 minutes. Serve warm with ice cream or whipped topping.

Rhubarb Cake

Makes 10 to 12 servings.

½ cup	butter (room temperature)	125 mL
1½ cups	sugar	375 mL
1	egg	1
1 tsp.	vanilla	5 mL
¼ tsp.	salt	1 mL
2 cups	flour	500 mL
1 tsp.	baking soda	5 mL
2 cups	rhubarb, chopped	500 mL
1 cup	buttermilk*	250 mL
¼ cup	butter	60 mL
2 tsp.	cinnamon	10 mL
1 cup	brown sugar	250 mL

Preheat oven to 350°F (180°C). Grease a 9 x 13-inch (22.5 x 32.5 cm/3 L) baking pan. Cream together ½ cup (125 mL) butter and sugar; blend in egg and vanilla. In a separate bowl, sift together flour, baking soda, and salt. Take out 2 tbsp. (30 mL) of flour mix and blend with rhubarb. Add flour mixture to creamed butter in 3 parts, alternating with buttermilk. Gently stir in floured rhubarb. Pour into prepared pan. Blend ¼ cup (60 mL) butter, cinnamon, and brown sugar; spread evenly over batter. Bake for 45 minutes or until toothpick comes out clean.

****Substitution for buttermilk:** Place 1 tbsp. (15 mL) lemon juice or vinegar in a 1-cup (250-mL) measure and fill with milk. Let set 5 minutes.*

Did you know?

When you pick rhubarb, pull the leaf stalk cleanly from the crown. Don't cut it! Cutting the stalks leaves the plant vulnerable to diseases and pests.

Rhubarb Cake with Crumb Topping

Makes 10 to 12 servings.

2 cups	flour	500 mL
½ cup	butter, well rounded	125 mL
¼ tsp.	salt	1 mL
1 tsp.	baking powder	5 mL
1	egg	1
4 cups	rhubarb, cooked	1 L
1½ cups	sugar	375 mL
½ cup	flour	125 mL
½ cup	butter, melted	125 mL
2	eggs, beaten	2
	sugar	
	cinnamon	

Preheat oven to 350°F (180°C). Grease and flour 9 x 12-inch (22.5 x 30-cm) baking pan. Crumble 2 cups (500 mL) flour, butter, salt, and baking powder. Mix in egg with fork. Set aside ¾ cup (180 mL) of this mixture for topping; press remainder into prepared pan. Mix rhubarb, sugar, ½ cup (125 mL) flour, melted butter, and 2 eggs and put on top of crushed crumbs in pan. Cover with remaining crumbs. Sprinkle with sugar and cinnamon. Bake for 45 to 55 minutes.

Easy Rhubarb Cake

Makes 8 to 10 servings.

1½ cups	brown sugar	375 mL
½ cup	butter or shortening	125 mL
1	egg	1
1 cup	buttermilk	250 mL
1 tsp.	baking soda	5 mL
1 tsp.	vanilla	5 mL
½ tsp.	salt	2 mL
2 cups	flour	500 mL
1½ cups	rhubarb, finely chopped	375 mL
½ cup	sugar	125 mL
½ tsp.	cinnamon	2 mL

Preheat oven to 375°F (190°C). Grease and flour a 9 x 12-inch (22.5 x 30-cm) cake pan. Cream sugar and butter. Add egg, then buttermilk, baking soda, vanilla, salt, and flour, in order. Add cut-up rhubarb last. Pour batter into prepared pan. Mix sugar and cinnamon together and sprinkle on top. Bake for 30 to 35 minutes.

***Substitution for buttermilk:** Place 1 tbsp. (15 mL) lemon juice or vinegar in a 1-cup (250-mL) measure and fill with milk. Let set 5 minutes.*

Did you know?

Unless you are growing rhubarb for its ornamental value, cut off the seed stalks that shoot up alongside the leaf stalks. The seed stalks drain energy from the plant's leaf and petiole production, reducing your food crop. Seed stalk development tends to be more frequent when the spring is cool.

Lunar Rhubarb Cake

Makes 8 to 10 servings.

½ cup	butter	125 mL
1½ cups	sugar	375 mL
1	egg	1
1 tsp.	vanilla	5 mL
2 cups	rhubarb	500 mL
2 cups	flour	500 mL
1 tsp.	baking soda	5 mL
½ tsp.	salt	2 mL
1 cup	buttermilk*	250 mL
¼ cup	butter, softened	60 mL
2 tsp.	cinnamon	10 mL
1 cup	brown sugar, packed	250 mL

Preheat oven to 350°F (180°C). Grease and flour a 9x13-inch (22.5x32.5-cm) cake pan and set aside. Beat together ½ cup (125 mL) butter and sugar until creamy. Blend in egg and vanilla. Cut rhubarb into ½-inch (12-mm) pieces. Stir together flour, soda, and salt. Use a little of this mixture to dredge the rhubarb. Add buttermilk and dry ingredients to creamed mixture in 3 parts, alternating. Gently stir in the rhubarb. Put into the pan and smooth the top. Blend together ¼ cup (60 mL) butter, cinnamon, and brown sugar; sprinkle evenly over batter. Bake for 45 minutes or until cake comes away from the sides of the pan and the top resembles the surface of the moon.

Substitution for buttermilk: *Place 1 tbsp. (15 mL) lemon juice or vinegar in a 1-cup (250-mL) measure and fill with milk. Let set 5 minutes.*

Jeannie's Rhubarb Cake

Makes 8 to 10 servings.

1½ cups	brown sugar	375 mL
½ cup	butter	125 mL
2	eggs	2
1 cup	sour milk*	250 mL
1 tsp.	baking soda	5 mL
1 tsp.	salt	5 mL
1¼ cups	flour	310 mL
1 tsp.	vanilla	5 mL
2 cups	rhubarb, finely chopped	500 mL
½ cup	brown sugar	125 mL
1 tsp.	cinnamon	5 mL

Preheat oven to 350°F (180°C). Grease and flour 9 x13-inch (22.5x32.5-cm) pan. Cream 1½ cups (375 mL) brown sugar and butter. Beat in eggs until fluffy. Add sour milk, baking soda, salt, and flour; mix well. Add vanilla and rhubarb. Mix and pour into prepared pan. • In a small bowl, mix ½ cup (125 mL) brown sugar and cinnamon. Sprinkle on top of batter and bake for 45 minutes.

***Substitution for sour milk:** Place 1 tbsp. (15 mL) lemon juice or vinegar in a 1-cup (250-mL) measure and fill with milk. Let set 5 minutes.*

Did you know?

In theory, you can harvest all the leaf stalks from your rhubarb plant without damaging the health of the crown. However, research says there is some benefit to leaving the lowest-growing leaf stalks attached, allowing them to collect energy for the crown until new stalks grow. Smaller harvests let the crown build up strength for future growth.

Fresh Rhubarb Cake

Makes 8 servings.

½ cup	shortening	125 mL
¾ cup	brown sugar	180 mL
¾ cup	honey	180 mL
2	eggs	2
1 tsp.	vanilla	5 mL
2 cups	flour	500 mL
½ tsp.	salt	2 mL
1 tsp.	baking soda	5 mL
1 cup	sour milk or buttermilk*	250 mL
1 to 2 cups	rhubarb, finely diced	250 to 500 mL
½ cup	brown sugar	125 mL
1 tsp.	cinnamon	5 mL

Preheat oven to 350°F (180°C). Prepare 9 x 13-inch (22.5 x 32.5-cm) pan. Cream shortening. Add ¾ cup (180 mL) brown sugar, honey, eggs, and vanilla; beat well. Combine flour, salt, and baking soda; add to creamed mixture, alternating with sour milk. Mix until well blended. Fold in rhubarb. Pour into prepared pan. In a small bowl, combine ½ cup (125 mL) brown sugar and cinnamon. Sprinkle over batter. Bake for 35 to 40 minutes.

***Substitution for sour milk or buttermilk:** Place 1 tbsp. (15 mL) lemon juice or vinegar in a 1-cup (250-mL) measure and fill with milk. Let set 5 minutes.*

Rhubarb Bundt Cake

Makes 12 or more servings.

1½ cups	rhubarb (about 3 stalks)	375 mL
2 tbsp.	sugar	30 mL
2 tsp.	cinnamon	10 mL
½ cup	butter, softened	125 mL
2 cups	sugar	500 mL
4	eggs	4
½ cup	light vegetable oil	125 mL
3 tbsp.	lemon juice	45 mL
2 tsp.	vanilla	10 mL
3 cups	flour	750 mL
1 tbsp.	baking powder	15 mL
½ tsp.	salt	2 mL
½ cup	milk	125 mL
1 cup	pecans or walnuts, chopped	250 mL
	icing sugar	
	cinnamon, to taste	

Preheat oven to 350°F (180°C). Generously grease a 3-quart (3-L) bundt pan. Combine rhubarb with 2 tbsp. (30 mL) sugar and cinnamon; set aside. In a large mixing bowl, cream together butter and 2 cups (500 mL) sugar. Beat in eggs one at a time, then oil, lemon juice, and vanilla. In a separate bowl, sift together flour, baking powder, and salt. Add flour mixture to the creamed mixture in 3 parts, alternating with the milk. Fold in nuts. Spoon a third of batter into prepared pan. Top with half the rhubarb mixture. Repeat, then cover with remaining third of batter. Bake 60 to 65 minutes, or until an inserted skewer comes out clean. Set cake on cooling rack. When cool enough to handle, invert onto serving plate. Before serving, dust top with powdered sugar and a sprinkling of cinnamon.

Hint: *This cake freezes beautifully.*

Sandi's Rhubarb Cake

Makes 8 servings.

1½ cups	brown sugar	375 mL
¼ cup	white sugar	60 mL
½ cup	shortening	125 mL
1	egg	1
1 tsp.	vanilla	5 mL
1 cup	sour cream	250 mL
2 cups	flour	500 mL
1 tsp.	baking soda	5 mL
½ tsp.	salt	2 mL
1½ cups	rhubarb	375 mL
¾ cup	sugar	180 mL
1½ tsp.	butter	7 mL
1½ tsp.	cinnamon	7 mL
1 cup	nuts	250 mL

Preheat oven to 350°F (180°C). Grease and flour 9 x 13-inch (22.5 x 32.5 cm) cake pan. Cream together both sugars, shortening, egg, vanilla, and sour cream. In a separate bowl, sift together the flour, baking soda, and salt. Add to creamed mixture. Stir in the rhubarb. Pour into prepared pan. • In a small bowl, mix ¾ cup (180 mL) sugar, butter, cinnamon, and nuts together. Sprinkle over the cake batter and bake for 30 to 40 minutes. This is very good served with a strawberry sauce.

Alice's Rhubarb-and-Raspberry Cake

Makes 8 servings.

2 cups	flour	500 mL
½ cup	sugar	125 mL
3 tsp.	baking powder	15 mL
1 tsp.	salt	5 mL
2 tbsp.	margarine	30 mL
1 cup	milk	250 mL
1	egg	1
4 to 5 cups	fresh rhubarb, diced	1 to 1.25 L
3-oz. package	raspberry-flavoured gelatin	85-g package
½ to 1 cup	sugar	125 to 250 mL
½ to 1 cup	flour	125 to 250 mL
½ tsp.	cinnamon	2 mL
½ tsp.	nutmeg	2 mL
3 to 4 tbsp.	margarine	45 to 60 mL
½ cup	nuts, chopped (optional)	125 mL
	fresh strawberries (optional)	

Preheat oven to 350°F (180°C). Grease an 8 x 12-inch (20.5 x 30-cm) pan. Mix 2 cups (500 mL) flour, ½ cup (125 mL) sugar, baking powder, and salt. Work in 2 tbsp. (30 mL) margarine. Add milk and egg. Spread in pan. Spread rhubarb over bottom layer in pan and sprinkle with gelatin powder. • Mix ½ to 1 cup (125 to 250 mL) sugar, ½ to 1 cup (125 to 250 mL) flour, cinnamon, and nutmeg. Cut in 3 to 4 tbsp. (45 to 60 mL) margarine. Add nuts, if desired, then crumble mixture over gelatin and rhubarb. Bake on middle oven rack for 35 to 45 minutes until golden-brown and rhubarb is bubbly. After cooling, garnish with fresh strawberries, if desired.

Hint: *For a fun change, try cherry or strawberry-flavoured gelatin.*

Rhubarb Streusel Muffins

Makes 24 muffins.

3/4 cup	pecans	180 mL
1/4 cup	flour	60 mL
1/4 cup	sugar	60 mL
1/4 cup	butter, chilled	60 mL
3	eggs	3
1 cup	vegetable oil	250 mL
1 1/2 cups	brown sugar	375 mL
1 1/2 tsp.	vanilla	7 mL
2 1/2 cups	rhubarb, finely diced	625 mL
3 cups	flour	750 mL
2 tsp.	baking soda	10 mL
1 1/2 tsp.	cinnamon	7 mL
1/2 tsp.	salt	2 mL
1/2 tsp.	baking powder	2 mL
1/2 tsp.	nutmeg	2 mL
1/2 tsp.	allspice	2 mL

Preheat oven to 350°F (180°C). In food processor, process pecans until finely chopped; reserve. In processor combine 1/4 cup (60 mL) flour, sugar, and butter until crumbly. Mix in 1/4 cup (60 mL) processed pecans, setting remaining amount aside. Set topping aside. • In large bowl, use an electric mixer to beat eggs, oil, sugar, and vanilla until thick and foamy. Stir in rhubarb and reserved pecans. In separate large bowl, combine 3 cups (750 mL) flour, soda, cinnamon, salt, baking powder, nutmeg, and allspice. Gradually stir dry ingredients into rhubarb mixture, stirring just until no dry spots remain. Spoon batter into 24 greased or paper-lined muffin cups. Sprinkle topping evenly over each muffin. Bake 25 to 30 minutes or until cake tester comes out clean. Let muffins cool in pan 2 minutes. Remove from pan and let cool completely on wire rack.

Strawberry-Rhubarb Muffins

Makes 12 muffins.

1 cup	rolled oats	250 mL
1 cup	strawberry yogurt	250 mL
½ cup	oil	125 mL
¾ cup	brown sugar	180 mL
1	egg	1
1 cup	flour	250 mL
1 tsp.	salt	5 mL
½ tsp.	baking soda	2 mL
1 tsp.	baking powder	5 mL
1 tsp.	cinnamon	5 mL
1 cup	bran	250 mL
¾ cup	rhubarb, chopped	180 mL
¼ cup	sugar	60 mL
½ cup	strawberries, sliced	125 mL

Preheat oven to 375°F (190°C). Soak oats in yogurt in a large bowl. Add oil, brown sugar, and egg; beat well. Sift in flour, salt, baking soda, baking powder, cinnamon, and bran. Toss rhubarb in sugar. Add rhubarb and strawberries to oat mixture. Fill prepared muffin cups ⅔ full and bake for 20 minutes.

Mrs. O'Brien's Rhubarb Muffins

Makes 12 muffins.

½ cup	butter or margarine	125 mL
½ cup	sugar	125 mL
2	eggs	2
¾ cup	rhubarb jam	180 mL
¾ cup	flour	180 mL
¼ cup	whole-wheat flour	60 mL
1 tsp.	baking powder	5 mL
½ tsp.	baking soda	2 mL
¼ tsp.	salt	1 mL
½ tsp.	cinnamon	2 mL
1 cup	rolled oats	250 mL
¾ cup	brown sugar	180 mL

Preheat oven to 350°F (180°C). Cream together butter, sugar, and eggs. Blend in rhubarb jam. In a separate large bowl, mix flours, baking powder, baking soda, salt, cinnamon, rolled oats, and brown sugar. Make well in center; add creamed mixture. Stir until moistened. Fill greased and floured muffin cups ⅔ full. Bake for 20 to 25 minutes.

Did you know?

Be sure to rid your garden of perennial weeds before you plant rhubarb. Annual weeds can be controlled with a hoe or by hand. Check for weeds hiding under the canopy of the leaves!

Rhubarb Muffins

Makes 12 large muffins.

1¼ cups	brown sugar	310 mL
½ cup	oil	125 mL
1	egg	1
2 tsp.	vanilla	10 mL
1 cup	buttermilk*	250 mL
2½ cups	flour	625 mL
1 tsp.	baking powder	5 mL
1 tsp.	baking soda	5 mL
½ tsp.	salt	2 mL
1½ cups	rhubarb, diced	375 mL
½ cup	nuts, chopped	125 mL
1 tsp.	cinnamon	5 mL
⅓ cup	sugar	80 mL
1 tbsp.	butter or margarine, softened	15 mL

Preheat oven to 400°F (200°C). Cream together brown sugar and oil; add egg. Slowly stir in vanilla and buttermilk. In a separate bowl, combine flour, baking powder, baking soda, and salt. Mix well with creamed ingredients. Stir in rhubarb and nuts. Fill prepared muffin cups ⅔ full. • In a small bowl, mix together cinnamon, sugar, and butter or margarine; sprinkle over muffins. Bake for 20 to 25 minutes.

***Substitution for buttermilk:** Place 1 tbsp. (15 mL) lemon juice or vinegar into a 1-cup (250-mL) measure and fill with milk. Let set 5 minutes.*

Phantom Rhubarb Muffins

Makes 12 large muffins.

½ cup	sour cream	125 mL
¼ cup	vegetable oil	60 mL
1	egg	1
1⅓ cups	flour	330 mL
1 cup	rhubarb, diced	250 mL
⅔ cup	brown sugar	160 mL
½ tsp.	baking soda	2 mL
¼ tsp.	salt	1 mL
¼ cup	brown sugar	60 mL
¼ cup	chopped nuts	60 mL
½ tsp.	cinnamon	2 mL
2 tsp.	butter, melted	10 mL

Preheat oven to 375°F (190°C). Blend together sour cream, oil, and egg; set aside. In another bowl, stir together flour, rhubarb, ⅔ cup (160 mL) brown sugar, baking soda, and salt; combine with wet ingredients. Mix just until moistened. Fill 12 large muffin cups ⅔ full. • Combine ¼ cup (60 mL) brown sugar, nuts, cinnamon, and butter; spoon small amount on each muffin. Bake for 25 to 30 minutes.

Did you know?

Although rhubarb is a perennial, it won't last forever. If left alone, the crown will grow outward each year, leaving the centre tissues to die and rot. A well-tended rhubarb plant will live up to 15 years; but most gardeners should replace their plants every 7 to 10 years.

Rhubarb Nut Bread

Makes 2 loaves.

1½ cups	brown sugar	375 mL
⅔ cup	vegetable oil	160 mL
1	egg	1
1 cup	sour milk*	250 mL
1 tsp.	baking soda	5 mL
1 tsp.	vanilla	5 mL
1 tsp.	salt	5 mL
2½ cups	flour	625 mL
1½ cups	rhubarb, diced	375 mL
½ cup	walnuts or pecans, chopped	125 mL
⅓ cup	sugar	80 mL
1 tbsp.	butter or margarine, melted	15 mL

Preheat oven to 325°F (160°C). Lightly grease and flour 2 9-x-5-inch (23.5 x 12.5-cm) loaf pans. Combine brown sugar, vegetable oil, and egg in mixing bowl. In separate bowl, combine sour milk, baking soda, vanilla, and salt. Alternating with flour, add milk mixture to sugar mixture, beating well after each addition. Fold in rhubarb and nuts. Place in prepared pans. In a small bowl, combine sugar and melted butter or margarine; sprinkle on loaves Bake for 45 minutes. Cool on wire racks.

***Substitution for sour milk:** Place 1 tbsp. (15 mL) lemon juice or vinegar in a 1-cup (250-mL) measure and fill with milk. Let set 5 minutes.*

Rhubarb Quick Bread

Makes 2 loaves.

3	eggs	3
1 cup	oil	250 mL
2 cups	brown sugar, packed	500 mL
2 tsp.	vanilla	10 mL
2½ cups	rhubarb, finely diced	625 mL
½ cup	walnuts, chopped	125 mL
3 cups	flour	750 mL
2 tsp.	baking soda	10 mL
2 tsp.	ground cinnamon	10 mL
1 tsp.	salt	5 mL
½ tsp.	baking powder	2 mL
½ tsp.	ground nutmeg	2 mL
½ tsp.	allspice	2 mL

Preheat oven to 350°F (180°C). Grease 2 9-x-5-inch (23.5 x 12.5-cm) loaf pans. Combine eggs, oil, brown sugar, and vanilla; beat with mixer until thick and foamy. Stir in rhubarb and walnuts. In a separate bowl combine flour, baking soda, cinnamon, salt, baking powder, nutmeg, and allspice. Add dry ingredients to the rhubarb mixture and stir just until blended. Divide batter between 2 prepared pans. Bake for 70 minutes or until cooked through. Cool in pans, then turn onto rack.

Did you know?

Field tests in Brooks, Alberta found that the Mcdonald variety of rhubarb produces nicely coloured leaf stalks on vigorous plants and makes great pie fillings. Early Sunrise was another good choice, although less vigorous, and Valentine was recommended for its bright-red stalks and minimal seed-stalk growth. The Strawberry variety also performed well.

Rhubarb Cornbread

Makes 8 servings.

1½ cups	yellow cornmeal	375 mL
½ cup	flour	125 mL
2 tsp.	baking powder	10 mL
1 tsp.	salt	5 mL
3	eggs, well beaten	3
1¼ cup	milk	310 mL
⅓ cup	unsalted butter, melted	80 mL
¼ cup	honey	60 mL
½ cup	currants, dried	125 mL
3	garlic cloves, minced	3
1 tsp.	tangerine zest	5 mL
1 cup	corn, frozen	250 mL
¼ cup	cilantro, minced	60 mL
¼ cup	rhubarb, chopped	60 mL
¼ cup	pine nuts	60 mL
	olive oil	

Preheat oven to 400°F (200°C). Grease and flour a round 10-inch (25-cm) baking dish. Place the cornmeal, flour, baking powder, and salt in a large mixing bowl; mix well. In a separate bowl, combine eggs, milk, butter, and honey. Add currants, garlic, tangerine zest, corn, cilantro, and rhubarb. Toast pine nuts in olive oil and add to mixture. Mix just until dry ingredients are moistened, leaving plenty of lumps. Pour batter into prepared baking dish and bake for about 30 minutes. The cornbread is done when a knife pushed deep into the centre comes out clean.

Rhubarb Tea Bread

Makes 2 loaves.

3	eggs	3
1 cup	salad oil	250 mL
2 cups	brown sugar	500 mL
2 tsp.	vanilla	10 mL
2½ cups	rhubarb, diced	625 mL
½ cup	walnuts, chopped	125 mL
3 cups	flour	750 mL
2 tsp.	baking soda	10 mL
1 tsp.	salt	5 mL
½ tsp.	baking powder	2 mL
½ tsp.	ground nutmeg	2 mL
½ tsp.	ground allspice	2 mL
2 tsp.	cinnamon	10 mL

Preheat oven to 375°F (190°C). Grease 2 9-x-5-inch (23.5 x 12.5-cm) loaf pans. In a large bowl, beat together eggs, oil, sugar, and vanilla until thick and foamy. Stir in rhubarb and nuts. In a separate bowl, combine flour with baking soda, salt, baking powder, and spices. Stir until thoroughly blended. Add dry ingredients to rhubarb mixture, stirring gently until just blended. Spoon batter into prepared pans. Bake for one hour or until bread begins to pull away from sides of pans and a wooden skewer inserted in centre comes out clean. Let cool in pan for 10 minutes, then turn out onto a rack to cool completely.

Rhubarb Bread

Makes 2 loaves.

1	egg	1
1 cup	honey	250 mL
½ cup	butter, melted	125 mL
½ cup	pineapple **or** orange juice	125 mL
1 cup	rhubarb, finely chopped	250 mL
½ cup	nuts, chopped	125 mL
2½ cups	flour	625 mL
2 tsp.	baking powder	10 mL
½ tsp.	baking soda	2 mL
½ tsp.	salt	2 mL
¼ tsp.	ground ginger	1 mL

Preheat oven to 350°F (180 °C). Grease 2 smaller-sized loaf pans. In a mixing bowl, beat the egg with the honey, melted butter, and pineapple or orange juice. Stir in the rhubarb and nuts. In a separate bowl, sift together the flour, baking powder, baking soda, salt, and ginger. Combine dry ingredients with rhubarb mixture, stirring just to mix. Pour the batter into prepared pans. Bake 35 to 40 minutes or until the tops feel springy to the touch. Cool bread in pans 10 minutes before removing to cool on a rack.

Rhubarb Roll

Makes 6 servings.

2 cups	flour	500 mL
4 tsp.	baking powder	20 mL
pinch	salt	pinch
4 tbsp.	shortening or margarine	60 mL
	milk	
3 cups	rhubarb, diced	750 mL
1 cup	sugar	250 mL
2 tbsp.	butter or margarine	30 mL
1 tsp.	cinnamon	5 mL
1½ cups	brown sugar	375 mL
1¼ cup	water	310 mL
1 tbsp.	butter or margarine	15 mL

Preheat oven to 350°F (180°C). Prepare baking pan. Mix flour, baking powder, and salt together. Cut in shortening. Add just enough milk to make a soft dough. Roll out dough in large rectangle about ⅜ inch (1 cm) thick. Spread rhubarb and sugar on dough. Dab with 2 tbsp. (30 mL) butter and sprinkle with cinnamon. Roll up and place in pan. Mix brown sugar, water, and 1 tbsp. (15 mL) butter and heat to melt butter. Pour over roll. Bake for 50 minutes.

Rhubarb Rolls

Makes 8 servings.

1½ cups	sugar	375 mL
1½ cups	water	375 mL
3 cups	flour	750 mL
3 tsp.	baking powder	15 mL
⅓ cup	sugar	80 mL
½ tsp.	salt	2 mL
½ cup	shortening or margarine	125 mL
1 cup	milk	250 mL
3 tbsp.	margarine, melted	45 mL
3 cups	rhubarb, diced	750 mL
1 cup	rhubarb, diced	250 mL
½ cup	sugar	125 mL
⅔ cup	water	160 mL
1 cup	whipping cream	250 mL

Preheat oven to 375°F (190°C). Grease a 9 x 13 x 2-inch (22.5 x 33.5 x 5-cm) baking dish. Combine 1½ cups (375 mL) sugar and 1½ cups (375 mL) water, bring to boil, and simmer for 5 minutes. Pour into prepared baking dish and set aside. • Mix flour, baking powder, sugar, and salt together. Cut in shortening. Stir in milk until flour is moistened. Knead on lightly floured board about 20 seconds. Roll to a 12-inch (30-cm) square. Brush with melted margarine. Spread 3 cups (750 mL) rhubarb on dough. Roll up as jellyroll; cut into 1½-inch (3.5-cm) slices and place evenly in the syrup in the baking dish. Bake for 35 to 40 minutes. • While rolls are baking, cook remaining 1 cup (250 mL) rhubarb, ½ cup (125 mL) sugar, and ⅔ cup (160 mL) water until rhubarb is soft and sauce is thickened. Baste rolls in oven with this sauce during last 20 minutes of baking time. Serve rolls hot in sauce dishes with extra cream.

Rhubarb Cream Cheese Coffee Cake

Makes 10 to 12 servings.

1¼ cups	rhubarb, finely diced	310 mL
½ cup	sugar	125 mL
1	fresh lemon slice	1
2¼ cups	flour	560 mL
¾ cup	sugar	180 mL
¾ cup	butter	180 mL
½ cup	walnuts, chopped	125 mL
½ tsp.	cinnamon	2 mL
¼ tsp.	salt	1 mL
½ tsp.	baking soda	2 mL
½ tsp.	baking powder	2 mL
¾ cup	sour cream	180 mL
1 tsp.	vanilla	5 mL
2	eggs	2
1 tsp.	lemon zest	5 mL
8-oz. package	cream cheese, softened	250-g package
¼ cup	sugar	60 mL

Preheat oven to 350°F (180°C). Grease a 10-inch (25-cm) springform pan. In a saucepan, combine rhubarb, sugar, and lemon slice; bring to a boil over high heat, stirring until sugar melts. Decrease heat; simmer for 12 to 15 minutes or until thickened and reduced to ½ cup (125 mL). Cool. • In a large bowl, combine flour and sugar. Cut in butter until mixture resembles coarse crumbs. Place walnuts and cinnamon into 1 cup (250 mL) of flour mixture and aside for topping. Add salt, baking soda, and baking powder to remaining flour mixture. In a small bowl, stir together sour cream, vanilla, and 1 egg. Stir sour cream mixture into flour mixture just until blended. With floured fingers, press sour cream mixture evenly over bottom and 2 inches (5 cm) up sides of prepared pan. Mix lemon zest, cream cheese, sugar, and remaining egg. Layer over sour cream mixture. Top with rhubarb mixture. Sprinkle topping over rhubarb. Bake for 1 hour or until knife comes out clean. Serve warm.

Rhubarb Coffee Cake

Makes 15 servings.

2 cups	flour	500 mL
1 cup	sugar	250 mL
1 tsp.	baking powder	5 mL
1 tsp.	baking soda	5 mL
½ tsp.	salt	2 mL
4 tbsp.	butter or margarine	60 mL
2	large eggs	2
1 cup	buttermilk*	250 mL
2 tsp.	vanilla	10 mL
4 cups	rhubarb	1 L
¾ cup	flour	180 mL
½ cup	brown sugar, packed	125 mL
6 tbsp.	butter or margarine	90 mL
1 tsp.	ground cinnamon	5 mL

Preheat oven to 375°F (190°C). Grease and flour 9 x 13-inch (22.5 x 32.5 cm/3 L) baking pan. In large bowl, combine 2 cups (500 mL) flour, sugar, baking powder, baking soda, and salt. With pastry blender or 2 knives, cut in 4 tbsp. (60 mL) butter until mixture resembles large crumbs. In small bowl, beat eggs with buttermilk and vanilla; stir into flour mixture until moistened. Spoon batter into baking pan. Cut rhubarb into 1-inch (2.5-cm) pieces and place evenly over batter in baking pan. • Blend ¾ cup (180 mL) flour, brown sugar, 6 tbsp. (90 mL) butter (slightly softened), and cinnamon until mixture comes together. Scatter over rhubarb and bake for 45 minutes or until toothpick inserted in centre comes out clean. Cool cake in pan on wire rack 15 minutes to serve warm or cool to serve later.

***Substitution for buttermilk:** Place 1 tbsp. (15 mL) lemon juice or vinegar in a 1-cup (250-mL) measure and fill with milk. Let set 5 minutes.*

Rhubarb Squares

Makes 9 squares.

3 cups	rhubarb	750 mL
3-oz. package	strawberry-flavoured gelatin	85-g package
1 cup	flour	250 mL
2 tbsp.	sugar	30 mL
½ tsp.	baking powder	2 mL
½ tsp.	salt	2 mL
¼ cup	oil	60 mL
¼ cup	milk	60 mL
1	egg	1
½ cup	sugar	125 mL
½ cup	flour	125 mL
¼ cup	butter or margarine	60 mL

Preheat oven to 350°F (180°C). Grease and flour 9-inch (22.5-cm) square baking pan. Combine rhubarb and gelatin; mix well and set aside. Combine 1 cup (250 mL) flour, 2 tbsp. (30 mL) sugar, baking powder, and salt; add oil and stir until mixture resembles coarse crumbs. Add milk and egg, mixing until smooth. Press mixture into bottom and ¾ inch (2 cm) up sides of pan. Spread rhubarb mixture over top and set aside. • Combine ½ cup (125 mL) sugar and flour; cut in butter until mixture is crumbly. Sprinkle over rhubarb mixture. Bake for 35 to 40 minutes. Cut into 3-inch (10-cm) squares. Serve with ice cream or whipped cream.

Rhubarb Spring Celebration

Makes 8 servings.

7 cups	rhubarb	1.75 L
1 cup	water	250 mL
1 cup	ginger snaps, crushed	250 mL
⅔ cup	sugar	160 mL
½ tsp.	cinnamon	2 mL
3 tbsp.	unsalted butter, melted	45 mL
1 pint	ice cream **or** whipping cream, whipped	500 mL

Preheat oven to 350°F (180°C). Grease a 1-quart (1-L) baking dish. In a saucepan, combine rhubarb with water. Cover and simmer 10 minutes. Drain well. Combine ginger snaps, sugar, cinnamon, and butter. Place a layer of rhubarb in bottom of dish, then a layer of cookie mixture. Repeat layering. Bake for 15 minutes. Serve warm with ice cream or whipped cream.

Rhubarb Delight

Makes 9 to 12 servings.

1 cup	flour	250 mL
1/4 tsp.	salt	1 mL
2 tbsp.	sugar	30 mL
1/2 cup	butter	125 mL
1 cup	sugar	250 mL
2 tbsp.	flour	30 mL
1/3 cup	cream	80 mL
3	egg yolks	3
2 1/4 cups	rhubarb, very finely chopped	560 mL
3	egg whites	3
1/4 tsp.	cream of tartar	1 mL
6 tbsp.	sugar	90 mL

Preheat oven to 325°F (160°C). Prepare an 8 x 8-inch (20 x 20-cm) baking pan. Stir together 1 cup (250 mL) flour, salt, and 2 tbsp. (30 mL) sugar. Cut in butter until crumbly; place mixture into pan. Bake for 10 minutes or until lightly browned. • Raise oven temperature to 400°F (200°C). Mix 1 cup (250 mL) sugar, 2 tbsp. (30 mL) flour, cream, egg yolks, and rhubarb in saucepan; cook over low heat until thick. Pour into crust. • Beat egg whites and cream of tartar until foamy. Gradually add the sugar, a spoonful at a time, beating between additions. Continue beating until meringue is thick and glossy and sugar is all dissolved. Meringue should be smooth and stand in stiff peaks. Spread meringue over the filling. Bake until lightly browned.

Matrimonial Bars

Makes 9 to 12 servings.

4 cups	rhubarb	1 L
1½ cups	sugar	375 mL
2 tbsp.	cornstarch	30 mL
1 tsp.	vanilla	5 mL
1 tsp.	lemon or orange zest	5 mL

Crust

1½ cups	rolled oats	375 mL
1½ cups	flour	375 mL
½ tsp.	baking soda	2 mL
1 tsp.	baking powder	5 mL
¼ tsp.	salt	1 mL
1 cup	brown sugar	250 mL
1 cup	butter	250 mL

Preheat oven to 350°F (180°C). Combine rhubarb, sugar, cornstarch, vanilla, and lemon/orange zest in saucepan; cook until thick. Cool completely. • Meanwhile, combine rolled oats, flour, baking soda, baking powder, salt, brown sugar, and butter. Pat ⅔ of this mixture into a greased 9 x 9-inch (22.5 x 22.5-cm) pan. Add filling and sprinkle with remaining crumbs. Bake for 30 to 35 minutes. Chill before cutting.

Variation
Use traditional date filling, below, adding juice after cooking.

½ lb.	chopped dates	230 g
½ cup	water	125 mL
2 tbsp.	brown sugar	30 mL
1 tsp.	orange zest	5 mL
2 tbsp.	orange juice	30 mL
1 tsp.	lemon zest	5 mL

Rhubarb Dessert

Makes 8 servings.

1	egg	1
2 cups	flour	500 mL
½ cup	butter, melted	125 mL
¼ tsp.	salt	1 mL
1 tsp.	baking powder	5 mL
1½ cups	sugar	375 mL
½ cup	butter, melted	125 mL
2	eggs, beaten	2
½ cup	flour	125 mL
4 cups	rhubarb, diced	1 L
1 tsp.	sugar	5 mL
	cinnamon, to taste	

Preheat oven to 375°F (190°C). Prepare an 8 x 8-inch (20 x 20-cm) pan. In a bowl, beat egg until light. Add 2 cups (500 mL) flour, ½ cup (125 mL) melted butter, salt, and baking powder. Mix with a fork until crumbly. Reserve one cupful for the top; press the remaining crumbs into prepared pan. • Combine sugar and ½ cup (125 mL) melted butter. Add 2 eggs, ½ cup (125 mL) flour, and rhubarb. Mix; pour over base. Sprinkle the remaining crumbs over filling, then sprinkle 1 tsp. (5 mL) sugar and cinnamon over the crumbs. Bake for 40 minutes.

Rhubarb Peekaboo

Makes 8 servings.

1 package	2-layer white cake mix	1 package
4 cups	rhubarb	1 L
1 cup	sugar	250 mL
1 tsp.	lemon zest	5 mL
	powdered sugar	

Preheat oven to 350°F (180°C). Prepare 7½ x 12-inch (18.75 x 30-cm) baking pan. Make cake as directed and pour into prepared baking pan. Top with diced rhubarb and lemon zest. Sprinkle sugar over all. Bake for 40 to 45 minutes. As the peekaboo bakes, the rhubarb sinks to the bottom, forming a sauce, and the cake rises to top. Sprinkle top of baked cake with powdered sugar. Serve warm with cream.

desserts

more than just pies

Other Rhubarbs

Rheum palmatum *(Chinese rhubarb, Turkey rhubarb, East Indian rhubarb) is native to northwest China and northeast Tibet, and is hardy only to zone 6, according to various Royal Horticultural Society publications, yet we have grown it at the Devonian Botanic Garden from seed. In fact, it thrives in a northern climate, so, as is the case of a great many herbaceous perennials, do not believe all you read about their hardiness. It is a rhizomatous perennial that develops a truly massive rootstock and thick petioles. These are not for eating, although there is at least one report of possible culinary use on the continent. The superintendent of exotics at Versailles, Monsieur Thouin, wrote Dr. Anthony Fothergill in 1785 about boiling the stems, stripped of their bark and strings, with an equal quantity of honey or sugar to make a marmalade, which was considered to be "a mild and pleasant laxative, and highly salubrious."*

Rheum palmatum *is one of the ornamental species. The dark-green leaves are broadly ovate to rounded, palmately 3 to 9-lobed, and coarsely toothed; they grow to about 90 centimetres in length. Their underside is red or purple-red and softly hairy. In the early summer, numerous tiny, star-shaped, creamy green to dark-red flowers are borne in panicles that grow to 2 metres in height. The cultivar "Atrosanguinium" (syn. "Atropurpureum") is even more spectacular as its leaves emerge from scarlet buds. These leaves are first a crimson-purple, then they fade to dark green above. Cerise-pink flowers are borne in panicles.*

Rheum palmatum *v.* tanguticum *(syn.* R. tanguticum*) bears dark-purple leaves that are less deeply cut than other species. Its flowers appear on erect flowering side shoots instead of being terminal, as in the other forms.*

Other species growing at the Devonian Botanic Garden include R. alexandrae, R. australis, R. palmatum *v.* tanguticum, *and* R. compactum *(dwarf rhubarb).* R. compactum *is a native of China and is one of the smaller, less-imposing ornamental rhubarbs. The leaves are quite glossy and heart-shaped with wavy margins. Panicles with many drooping branches bearing creamy-white flowers are produced in the summer.*

Rhubarb Stewed with Apples and Strawberries

Makes 8 ⅔-cup (160-mL) servings.

1	orange	1
4 cups	fresh or frozen rhubarb	1 L
1	large apple	1
1 cup	water	250 mL
¼ cup	sugar	60 mL
2 cups	fresh strawberries	500 mL
	sugar to taste	
½ cup	low-fat plain yogurt	125 mL
2 tbsp.	brown sugar, packed	30 mL

Grate rind and squeeze juice from orange. Cut rhubarb into 1-inch (2.5-cm) lengths. Peel, core, and thinly slice apple. In saucepan, combine orange zest and juice, rhubarb, apple, water, and ¼ cup (60 mL) sugar; cover and bring to boil. Reduce heat and simmer for 10 minutes or until fruit is tender, stirring occasionally. Remove from heat and stir in strawberries. Add sugar to taste. Top each serving with yogurt and sprinkle with brown sugar. Serve warm or at room temperature.

Baked Rhubarb

Makes 4 servings.

3 cups	rhubarb	750 mL
1 cup	honey	250 mL
2	egg whites	2

Preheat oven to 350°F (180°C). Prepare a glass baking dish. Cook rhubarb and honey until rhubarb is tender, about 10 minutes. Beat egg whites until stiff. In baking dish, fold half of egg whites into rhubarb. Spoon remaining egg white on top. Place in oven for about 15 minutes or until lightly browned.

Stewed Rhubarb

Makes about 2 cups (500 mL).

4 cups	rhubarb	1 L
½ cup	sugar	125 mL
½ cup	water	125 mL

Place all ingredients into a sauce pan and bring to boil. Reduce heat and simmer until rhubarb is very tender (the slower it cooks the better, as the rhubarb will retain its shape rather than becoming a mush). For a purée, drain off the juice and whisk the rhubarb to a smooth pulp. It will be soft enough to do it by hand. This dish is wonderful served on pancakes, crêpes, and ice-cream or by itself for breakfast.

Rhubarb Fruit Whip

Makes 6 to 8 servings.

2 cups	chopped rhubarb	500 mL
½ cup	sugar	125 mL
2 tbsp.	water	30 mL
1 tbsp.	gelatin	15 mL
¼ cup	cold water	60 mL
¼ cup	boiling water	60 mL
½ cup	sugar	125 mL
1 tsp.	lemon zest	5 mL
1 tbsp.	lemon juice	15 mL
1 tsp.	vanilla	5 mL
4	egg whites	4
½ cup	sugar	125 mL
⅛ tsp.	salt	0.5 mL

Place rhubarb, ½ cup (125 mL) sugar, and water in saucepan and cook for 5 minutes. Soak gelatin in cold water and then dissolve it in boiling water. Stir in sugar and lemon zest until sugar is dissolved. Add lemon juice, cooked rhubarb, and vanilla. Place mixing bowl with these ingredients in a container of ice water. When chilled, whip with an egg beater until frothy. In separate bowl, whip egg whites until stiff, adding ½ cup (125 mL) sugar gradually. Fold whites into gelatin mixture. Chill well before serving.

Rhubarb Scallop with Meringue

Makes 6 servings.

2 cups	rhubarb, diced	500 mL
1 cup	sugar	250 mL
1	orange, zest	1
1/4 tsp.	salt	1 mL
1	small sponge cake	1
2	egg whites	2
2 tbsp.	powdered sugar	30 mL

Preheat oven to 350°F (180°C). Prepare a baking dish. Cut rhubarb into 1-inch (25-mm) pieces. Add sugar, orange zest, and salt, mixing well. Cut sponge cake into thin slices; line bottom of baking dish with 3 or 4 slices. Cover layer with rhubarb. Continue layering cake and rhubarb until all is used. Cover and bake for 30 minutes. Beat egg whites until stiff; add powdered sugar slowly, beating until blended. Pile on baked mixture and bake 15 minutes longer, or until meringue is slightly browned.

Rhubarb Sherbet

Makes about 4 cups (1 L).

2 cups	rhubarb, diced	500 mL
1 cup	sugar	250 mL
2½ cups	water	625 mL
2 tbsp.	fresh lime juice	30 mL
	mint sprigs	
	lime slices	

Place the rhubarb and sugar into a saucepan with the water; bring to a boil and simmer, covered, until the sugar is dissolved and the rhubarb tender, about 5 minutes. Cool and purée. Add the lime juice and chill thoroughly. Make the sherbet in an ice-cream machine according to the manufacturer's directions. Garnish with mint and thin slices of lime.

Freezer Rhubarb Sherbet

Makes about 3 cups (750 mL).

2 cups	rhubarb, diced	500 mL
½ cup	sugar	125 mL
pinch	salt	pinch
1 cup	heavy cream	250 mL
2	egg yolks, beaten	2
1 tbsp.	lemon juice	15 mL
¼ tsp.	vanilla	1 mL
2	egg whites	2
¼ cup	sugar	60 mL

Place rhubarb, ½ cup (125 mL) sugar, and salt into saucepan; cover and simmer until rhubarb is tender. Cool. Combine cream, egg yolks, lemon juice, and vanilla. Mix with rhubarb. Freeze to mush in refrigerator tray at coldest setting. Beat egg whites, gradually adding ¼ cup (60 mL) sugar. Continue beating until stiff. Turn frozen mixture into chilled bowl and beat until smooth but not melted. Fold in egg white mixture and return to freezer. Freeze until firm.

Rhubarb and Ginger Mousse

Makes 4 servings.

4 cups	rhubarb	1 L
5 tbsp.	clear honey	75 mL
½	orange, juice and zest	½
¼ tsp.	ginger, ground	1 mL
2 tsp.	powered gelatin	10 mL
2 tbsp.	water	30 mL
2	egg whites	2

Chop rhubarb into 1-inch (2.5-cm) pieces and put into a pan with the honey, orange juice and zest, and ginger; simmer gently until the fruit is soft. Sprinkle gelatin over cold water and set aside for 3 minutes to soften. Add gradually to hot mixture and stir to dissolve. Cool the rhubarb mixture until it is half-set. Beat egg whites until stiff and fold them lightly into the rhubarb mixture. Spoon into decorative glasses and chill until set.

Saucy Rhubarb Dumplings

Makes 5 servings.

5 cups	rhubarb, diced	1250 mL
3/4 cup	sugar	180 mL
2 tbsp.	cornstarch	30 mL
1/2 cup	water	125 mL
2 tbsp.	butter or margarine	30 mL
1 cup	flour	250 mL
2 tsp.	baking powder	10 mL
1/2 tsp.	salt	2 mL
1 tsp.	sugar	5 mL
1/2 cup	milk	125 mL
2 tbsp.	butter or margarine	30 mL

Place rhubarb in a large, deep skillet or heavy pot. Mix 3/4 cup (180 mL) sugar and cornstarch together and add to rhubarb. Add water and butter. Bring to a boil, stirring gently for one minute. Reduce heat to low. • Make the dough by combining flour, baking powder, salt, and 1 tsp. (5 mL) sugar in a small bowl. Cut in butter. Add milk, stirring until just blended. Drop by small spoonfuls onto the simmering rhubarb. Cover and cook for 20 minutes, or until dough is cooked through. Serve warm with heavy cream.

Variation: *Use 3 cups (750 mL) of rhubarb plus 2 cups (500 mL) of sliced strawberries.*

Did you know?

Researchers have noticed that the more strongly red the stalks are, the less vigorous the plant's growth. Strongly red-stalked varieties also tend to be more susceptible to redleaf.

Rhubarb Mousse Gateaux

Makes 8 to 10 slices.

1	8 or 9-inch (20 or 23-cm) white sponge cake or plain white round cake	1
4 cups	fresh or frozen rhubarb, chopped	1 L
1/4 cup	water	60 mL
1/4 cup	sugar	60 mL
2 packages	unflavoured gelatin	2 packages
1/2 cup	cold water	125 mL
2 cups	whipping cream, whipped	500 mL
1/2 to 3/4 cup	Cointreau **or** other orange liqueur	125 to 180 mL
	orange segments or orange rind	
	chocolate fans	
	toasted, sliced almonds	

Simmer rhubarb in water with sugar in a pot for 10 to 15 minutes, until mixture is softened and gently boiling. Remove from stove and set aside. • In a separate pan, soak gelatin in cold water for 10 minutes, then warm gently on stovetop until gelatin mixture liquefies. Add gelatin mixture to the rhubarb, mix thoroughly, and refrigerate for a minimum of 4 hours or overnight. At this point mixture can be stored in refrigerator up to one week or it can be frozen. (A dash of red food colouring could be added into mixture). • Slice the cake into 3 layers. Blend cooled mixture with 1 cup (250 mL) of whipped cream and stir in Cointreau to taste. Cointreau can also be drizzled on cake sections. (If a non-alcoholic cake is desired, substitute the zest of one large orange; orange juice can be sprinkled on the cake sections.) Top the lower and middle layer of sponge cake with 1/2 to 1 inch (12 to 25 mm) of filling, stacking layers on top of each other. Using a spatula, coat entire cake with remaining 1 cup (250 mL) of whipped cream. Decorate top with orange segments and chocolate fans or thin strips of orange rind. Decorate sides of cake with toasted sliced almonds.

Rhubarb-Strawberry Sorbet

Makes 8 servings.

1½ cups	fresh rhubarb, chopped	375 mL
1 cup	sugar	250 mL
¾ cup	water	180 mL
2 cups	strawberries	500 mL
1 tbsp.	lemon juice	15 mL
1 tsp.	orange zest **or** 2 tbsp. (30 mL) orange liqueur	5 mL
¾ cup	water	180 mL

In saucepan, combine rhubarb, sugar, and ¾ cup (180 mL) water; simmer, covered, until rhubarb is very tender. Purée in food processor; transfer to bowl. In food processor, purée strawberries; stir into rhubarb mixture. Add lemon juice, orange zest or liqueur, and ¾ cup (180 mL) water. Freeze in an ice-cream maker following manufacturer's instructions. Alternatively, transfer to metal pan or bowl and freeze until barely firm. Process in food processor or beat with an electric mixer until smooth. Transfer to freezer container and freeze until firm. To serve, remove from freezer and let stand 15 to 30 minutes before serving or until mixture is soft enough to scoop. Serve on dessert plates surrounded with fresh berries or in sherbet glasses, each garnished with its own fruit or fresh mint leaf.

Rhubarb Sorbet

Makes 8 servings.

2/3 cup	sugar	160 mL
1/2 cup	water	125 mL
4 cups	rhubarb, fresh, diced	1 L
2 tsp.	lemon zest	10 mL
1 tbsp.	lemon juice	15 mL

Combine sugar and water in saucepan. Stir over low heat until sugar dissolves. Add remaining ingredients and continue cooking until rhubarb is tender, about 10 minutes. Blend mixture in processor or force through a sieve. Allow mixture to cool, then pour into ice cube tray. Place tray in freezer. Stir mixture with a fork several times during freezing process. Serve with whipped topping, if desired.

Strawberry-Rhubarb Sorbet

Makes 6 to 8 servings.

4 cups	rhubarb, fresh, diced	1 L
¼ cup	water	60 mL
½ to 1 cup	sugar	125 to 250 mL
1 pint	strawberries, fresh or frozen	500 mL

In a medium saucepan, combine rhubarb and water. Bring to a boil over medium heat. Cover and simmer until rhubarb is tender, about 5 minutes. Stir in sugar until dissolved. Purée rhubarb mixture in a blender until smooth. Pour into a large bowl; set aside. Process strawberries in blender until almost smooth. Stir into rhubarb mixture. Pour mixture into ice-cream maker and process according to manufacturer's instructions.

Rhubarb Cookies

Makes 3 to 4 dozen.

½ cup	butter	125 mL
1 cup	brown sugar	250 mL
1	egg	1
2 cups	white **or** whole-wheat flour	500 mL
2 tsp.	baking powder	10 mL
dash	salt	dash
½ tsp.	ground cloves	2 mL
½ tsp.	nutmeg	2 mL
1 tsp.	cinnamon	5 mL
¼ cup	milk	60 mL
½ cup	raisins	125 mL
1 cup	rhubarb, diced	250 mL
1 cup	walnuts, chopped	250 mL

Preheat oven to 350°F (180°C). Grease baking sheet. Cream together butter and sugar. Add the egg and beat until smooth. In a separate bowl, sift together the flour, baking powder, salt, cloves, nutmeg, and cinnamon. Add to the creamed mixture, alternating with the milk. Beat until smooth. Stir in the raisins, rhubarb, and walnuts. Drop by spoonfuls onto baking sheet and bake for 18 to 20 minutes, or until lightly browned and crisp. Remove to wire racks to cool completely.

Rhubarb and Strawberry Gratin

Makes 4 servings.

1¾ cups	strawberries, sliced	430 mL
2¼ cups	rhubarb (about 3 stalks)	560 mL
1 tbsp.	butter	15 mL
4 tbsp.	sugar	60 mL
2	egg yolks	2
2 tbsp.	sugar	30 mL
2 tbsp.	dry white wine	30 mL
½ cup	whipping cream, whipped	125 mL

Divide 1 cup (250 mL) of the strawberries evenly among 4 shallow, ovenproof dishes. Purée remaining strawberries in blender or food processor and set aside. Cut rhubarb into 1-inch (2.5-cm) pieces and sauté in butter with 4 tablespoons (60 mL) of sugar for 10 minutes or until tender. Cool and combine with strawberries in dishes. In a double boiler over boiling water, whisk egg yolks, sugar, strawberry purée, and wine for 4 to 6 minutes or until thick enough to coat a spoon. Add the whipped cream. Cover the fruit with the cream and place under a preheated broiler, about 4 inches (10 cm) from heat, until golden-brown and bubbly, about 30 seconds.

Rhubarb Fool

Makes 12 cups (3 L).

1	cinnamon stick	1
3	whole cloves	3
8 cups	rhubarb, chopped	2 L
1¼ cups	sugar	310 mL
2 cups	vanilla yogurt	500 mL
½ cup	whipping cream	125 mL

Place cinnamon and cloves in cheesecloth bag and tie. Combine rhubarb, sugar, and sachet of spices in saucepan and cook for 6 to 8 minutes. Put yogurt in a muslin bag (or use several layers of cheesecloth) in a colander and drain for 45 minutes. Whip cream and fold into yogurt. Place in serving bowls. Swirl cooked rhubarb into yogurt/cream mixture.

Tip: *Add red food colouring to enhance rhubarb color.*

Rhubarb Bavarian Dessert

Makes 9 to 12 servings.

1 cup	flour	250 mL
¼ tsp.	salt	1 mL
2 tsp.	sugar	10 mL
1/2 cup	margarine, softened	125 mL

Topping

4 cups	rhubarb	1 L
¼ cup	water	60 mL
1¼ cups	sugar	310 mL
dash	salt	dash
2 packages	unflavoured gelatin	2 packages
⅓ cup	cold water	80 mL
½ pint	whipping cream, whipped	250 mL

Preheat oven to 350°F (180°C). Grease a 9 x 9-inch (22.5 x 22.5-cm) pan. Combine flour, salt, and 2 tsp. (10 mL) sugar. Cut in margarine until mixture resembles coarse crumbs. Spread evenly in pan. Bake about 10 to 12 minutes until lightly browned. • Combine rhubarb, water, 1¼ cups (310 mL) sugar, and salt; simmer until rhubarb is tender, stirring occasionally. Soak gelatin in cold water. Add to rhubarb mixture and stir until dissolved. Chill until partially set; fold in whipped cream and pour on top of base. Chill for 3 hours before serving.

Baked Rhubarb and Jam

Makes 4 to 6 servings.

¼ cup	seedless red jam	60 mL
4 cups	rhubarb	1 L
½ tsp.	ground ginger	2 mL

Preheat oven to 350°F (180°C). Coat small baking pan with ⅓ of the jam. Cut rhubarb into 2-inch (5-cm) slices. Layer ½ the rhubarb on top of the jam layer. Sprinkle with ginger. Repeat jam and rhubarb layers. Bake covered for about 30 minutes.

Rhubarb Whip

Makes 8 servings.

3 cups	rhubarb sauce (see page 130 for simple sauce)	750 mL
6 tbsp.	sugar	90 mL
½ tsp.	vanilla	2 mL
½ tsp.	almond extract	2 mL
2 tbsp.	gelatin	30 mL
2 tbsp.	cool water	30 mL
½ cup	hot water	125 mL
2	egg whites	2
	whipped cream	

Place the rhubarb sauce in a bowl; add sugar, vanilla, and almond extract. Soak gelatin in 2 tbsp. (30 mL) of cool water for 5 minutes and then dissolve in hot water. Stir dissolved gelatin into rhubarb mixture. Beat egg whites stiff and fold into rhubarb mixture. Cover and chill. Serve in sherbet glasses with whipped cream, if desired.

Rhubarb Fritters

Makes 8 servings.

2	egg yolks	2
⅔ cup	milk	160 mL
1 tbsp.	butter, melted	15 mL
1 cup	flour	250 mL
1 tbsp.	sugar	15 mL
2	egg whites	2
2 cups	rhubarb, chopped	500 mL
1 cup	sugar	250 mL
	icing sugar	

Preheat enough cooking oil to cover fritters in deep fryer or deep skillet. Combine first 5 ingredients and blend well. Whip egg whites and fold into mixture. Dust rhubarb with sugar and fold into batter. Add a tablespoon of batter to oil to check temperature. Fritters should cook thoroughly in 5 to 7 minutes; adjust heat accordingly. Drain fritters on paper towels. Dust with icing sugar and serve.

Creamy Rhubarb Ring

Makes 12 servings.

4 cups	rhubarb, finely chopped	1 L
½ cup	sugar	125 mL
2 tbsp.	water	30 mL
3-oz. package	raspberry-flavoured gelatin	90-g package
10-oz. package	frozen raspberries	300-g package
1½ cups	vanilla ice cream	375 mL
2 tbsp.	cornstarch	30 mL
½ cup	whipping cream, whipped	125 mL

In large saucepan, combine rhubarb, sugar, and water; bring to a boil, stirring constantly. Reduce heat and cook until rhubarb is tender, about 15 minutes. Remove from heat; add gelatin, stirring until dissolved. Thaw raspberries and drain syrup; reserve syrup. Add raspberries and ice cream to rhubarb; stir until ice cream has melted and mixture is slightly thickened. Spoon into 6 ½-cup (125-mL) moulds. Refrigerate until firm. • In small saucepan, combine reserved syrup and cornstarch; blend well. Cook over low heat until mixture is clear and thickened, stirring constantly. Refrigerate until serving time. Just before serving, fold in whipped cream. Unmould rhubarb ring onto serving plate and serve with cream topping.

Rhubarb Soufflés with Citrus Strawberries

Makes 4 servings.

2 cups	rhubarb, fresh or frozen	500 mL
¼ cup	water	60 mL
1	egg, separated	1
1 tbsp.	cornflour	15 mL
2 tbsp.	castor sugar	30 mL
2	egg whites	2

Prepare recipe just before serving. *Preheat oven to 350°F (180°C). Cut rhubarb into 1-inch (2.5-cm) pieces and place in medium saucepan. Add water and bring to boil; reduce heat, cover, and simmer for about 10 minutes or until rhubarb is soft. Blend or process rhubarb mixture until smooth; push mixture through sieve, then return to saucepan. Separate egg; combine yolk with cornflour and sugar in small bowl and stir until smooth. Add cornflour mixture to rhubarb purée; stir over high heat until mixture boils and thickens. Remove from heat; cool 5 minutes. Add egg white to mixture, whisk until combined; transfer mixture to large bowl. • Beat 2 egg whites in small bowl until firm peaks form. Gently fold into rhubarb mixture in 2 batches. Pour mixture into 2 greased ovenproof soufflé dishes. Place dishes in small baking dish and pour in enough boiling water to come halfway up sides of dishes. Bake for 30 minutes. Serve soufflés immediately with citrus strawberries (recipe follows). Do not freeze or reheat in microwave.*

Citrus Strawberries

2 tbsp.	castor sugar	30 mL
¼ cup	water	60 mL
½ tsp.	orange zest	2 mL
¼ cup	orange juice	60 mL
2 cups	strawberries, halved	500 mL

Combine sugar and water in small saucepan; stir over high heat, without boiling, until sugar is dissolved. Bring to boil and boil uncovered, without stirring, for 4 minutes. Stir in zest and juice; cool mixture to room temperature. Add strawberries, cover, and refrigerate until cool. May be made ahead.

Cold Rhubarb Soufflé

Makes 4 servings.

4 cups	rhubarb, fresh, chopped	1 L
1 ½ cups	sugar	375 mL
	water	
1 package	unflavoured gelatin	1 package
2 tbsp.	crème de cassis (currant liqueur)	30 mL
1 cup	whipping cream	250 mL
4	egg whites	4
⅛ tsp.	cream of tartar	0.5 mL
½ cup	sugar	125 mL
1 tsp.	vanilla extract	5 mL

Combine rhubarb and sugar in saucepan. Add a small amount of water. Stew rhubarb until tender, about 10 minutes. Drain; reserve liquid. Put rhubarb through food mill or blender. Allow gelatin to dissolve in liqueur. Reduce reserved liquid by cooking to ½ cup (125 mL); cool and add to gelatin, then combine both with purée. Whip cream and set aside. Beat egg whites and cream of tartar. Add sugar and vanilla; beat until peaks form. Fold cream into rhubarb purée, followed by egg-white mixture. Make a collar of foil to extend at least 2 inches (5 cm) above edge of a 1½-quart (1.5-L) soufflé dish. Tie it around the container. Pour in soufflé mixture. Chill for 6 hours.

Did you know?

Rhubarb tolerates a wide range of soil pH and does well even in alkaline soil.

jams, jellies, & marmalades

more than just pies

Culinary Rhubarb

Rheum *x* cultorum *(syn.* R. rhabarbarum, R. rhaponticum, R. *x* hybridum*) is also known as garden rhubarb, pie plant, bastard rhubarb, English rhubarb, and sweet round-leaved dock. It is an ancient and complex hybrid, probably originally involving* R. rhaponticum *and* R. palmatum, *with many hybrids existing.*

*Culinary usage of rhubarb in Europe dates from the eighteenth century. Erasmus Darwin believed his "*R. hybridum*" was a mule rhubarb, perhaps being a serendipitous cross between* R. palmatum *and* R. rhaponticum, *both of which grew in his garden and those of his neighbours. It was he who in 1800 appears to have first drawn public attention to the use of the hybrid plant. These species and several others are found in China and other parts of the East.*

We have about fifty cultivars in the rhubarb collection at the Devonian Botanic Garden. These plants have large leaves and are grown for their thick leaf petioles. Petioles cut fresh and dipped in sugar make a satisfying, slightly acidic, thirst-quenching snack. Rhubarb petioles are also used for desserts, pie filling, sauces, jams, wine, and a myriad of other culinary uses.

Rhubarb-Ginger Preserves

Makes about 2 pints (1 L).

8 cups	rhubarb	2 L
4 cups	sugar	1 kg
4 tsp.	fresh ginger, chopped	20 mL

Cut rhubarb into 1-inch (2.5-cm) pieces and alternate layers of rhubarb and sugar in a large saucepan. Let stand 24 hours. Strain juice and reserve rhubarb. Add ginger to rhubarb juice. Bring to a boil, stirring constantly. Cook, without stirring, to 240°F (115°C) or until syrup, when dropped into cold water, forms a soft ball which flattens on removal from water (about 10 minutes). Add rhubarb and return mixture to boil; boil for 2 minutes. Pour into hot sterilized jars and seal. May be stored up to 6 months.

Rhubarb Jelly

Makes 8 half-pint (250-mL) jars.

4½ to 5 lbs.	rhubarb	2 to 2.25 kg
7 cups	sugar	1.75 kg
1 or 2 drops	red food colouring	1 or 2 drops
¾ cup	liquid fruit pectin	180 mL

Grind the rhubarb in a food processor or grinder. Strain through a jelly bag, reserving 3½ cups (875 mL) of juice. Pour juice into a large kettle; add sugar and food colouring if desired. Bring to a boil over high heat, stirring constantly. Add pectin; bring to a full rolling boil. Boil for 1 minute, stirring constantly. Remove from the heat; let stand a few minutes. Skim off foam. Pour into hot jars, leaving a ¼-inch (6-mm) headspace. Adjust caps. Process for 10 minutes in a boiling water bath.

Rhubarb-Ginger Jam

Makes 4 1-pint (500-mL) jars.

1 ½ lbs.	rhubarb, chopped	700 g
2 cups	sugar	500 mL
2-inch piece	peeled fresh ginger	5-cm piece
1 cup	water	250 mL

In a large saucepan, combine all ingredients. Simmer over medium-low heat, stirring occasionally and skimming foam as it rises, until rhubarb is soft and syrup begins to thicken, about 15 minutes. Remove rhubarb with a slotted spoon and set aside. Continue simmering syrup until very thick, another 7 to 10 minutes. Remove from heat. Discard ginger, return rhubarb to pan, stir, and cool for 5 to 10 minutes. Pour jam into 4 sterilized pint (500 mL) mason jars, seal, and let stand at room temperature to set. Store in refrigerator for up to 4 weeks.

Rhubarb and Ginger Jam

Makes 4 1-pint (500-mL) jars.

3 lbs.	rhubarb, chopped	1.4 kg
7 cups	sugar	1.75 kg
½ cup	preserved ginger, chopped	125 mL
6 tbsp.	lemon juice	90 mL

Combine rhubarb and sugar in a large saucepan; allow to stand 3 hours. Stir in ginger and lemon juice. Heat to boiling. Lower heat and simmer about 1 hour, until jam sheets from spoon. Skim off foam. Ladle hot jam into hot jars, filling to within ¼ inch (6 mm) of tops. Wipe rims and seal with new lids and rings. Process for 10 minutes in a boiling water bath. Remove jars from water bath and cool to room temperature. Check seals, label, and store in a cool, dry place.

Rhubarb and Grapefruit Jam

Makes 8 4-oz (125-mL) jars

1½ lbs.	rhubarb	700 g
2	grapefruit, large	2
3 cups	sugar	750 mL

Cut rhubarb into ½-inch (12-mm) pieces and place in a large bowl. Add zest from grapefruit. Pour the juice from the grapefruit and then the sugar over the rhubarb. Cover and leave overnight. • Slowly bring mixture to a boil in a heavy pan, stirring often. When the sugar has dissolved completely, raise the heat and boil hard for 15 minutes or until set. Pour into sterilized jars and seal.

Strawberry-Rhubarb Jam

Makes 5½ cups (1.3 L).

1½ cups	strawberries	375 mL
1½ cups	rhubarb, diced	375 mL
2½ cups	sugar	625 mL
1 cup	crushed pineapple, undrained	250 mL
3-oz. package	strawberry-flavoured gelatin	85-g package

In a large kettle, combine strawberries, rhubarb, sugar, and pineapple. Bring to a boil; reduce heat and simmer for 20 minutes. Remove from the heat; stir in gelatin until dissolved. Pour into jars or freezer containers, leaving a ½-inch (12-mm) headspace. Cool. Top with lids. Refrigerate or freeze.

Rhubarb-Blueberry Jam

Makes about 8 half-pint (250-mL) jars.

5 cups	fresh or frozen rhubarb, diced	1.25 L
1 cup	water	250 mL
5 cups	sugar	1.25 kg
21-oz. can	blueberry pie-filling	630-mL can
6-oz. package	raspberry-flavoured gelatin	180-g package

In a large kettle, combine rhubarb and water. Cook over medium-high heat for 4 minutes or until rhubarb is tender. Add sugar and bring to a boil; boil for 2 minutes. Stir in pie filling. Remove from the heat; cool for 10 minutes. Add gelatin and mix well. Place in jars, leaving a ¼-inch (6-mm) headspace. Process for 15 minutes in a boiling water bath.

Bluebarb Jam

Makes 5½ cups (1.3 L).

4 cups	rhubarb, diced	1 L
½ cup	water	125 mL
2¼ cups	frozen blueberries	560 mL
1 tbsp.	lemon juice	15 mL
1 box	regular powdered fruit pectin	1 box
5½ cups	sugar	1.4 kg

Place rhubarb and water in large, stainless-steel or enamel saucepan. Bring to a boil over high heat. Cover, reduce heat, and simmer 5 minutes, stirring frequently. Chop frozen blueberries coarsely and add to rhubarb with the lemon juice and pectin; mix well. Bring to a boil over high heat, stirring constantly. Add sugar, return to a full boil, and boil hard for 1 minute, stirring constantly. Remove from heat. Ladle jam into hot, sterilized jars, leaving a ¼-inch (6-mm) headspace. Seal immediately with 2-piece lids according to maker's instructions and process in boiling water bath for 5 minutes.

Ruby Jam

Makes 8 half-pint (250-mL) jars.

4 cups	rhubarb, diced	1 L
6 cups	sugar	1.5 kg
4 cups	fresh strawberries	1 L
1 cup	lemon juice	250 mL
1 cup	almonds, blanched, slivered	250 mL

Combine half of sugar with rhubarb and allow to stand for several hours. Crush berries, pour into kettle, and cover with remaining sugar. Add rhubarb and cook over low heat. Stir until sugar is dissolved. Increase heat and cook rapidly for 15 minutes. Add nuts and mix well. Continue cooking until jam coats the spoon. Pour into sterilized half-pint jars and seal.

Rhubarb-Pineapple Marmalade

Makes 4 half-pint (250-mL) jars.

3 cups	rhubarb	750 mL
1 cup	crushed pineapple	250 mL
2 cups	sugar	500 mL

Wash and peel rhubarb. Cut into ¼-inch (6-mm) pieces. Drain pineapple and combine with rhubarb in a large pot. Place on heat and cook, stirring constantly, until rhubarb is soft. Add sugar and bring to a boil. Lower heat and continue boiling; stir constantly to prevent burning. Cook for about 15 minutes. Pour into sterilized jars and seal with hot paraffin immediately.

Did you know?

Rhubarb is the one of the earliest garden fruits, emerging when soil temperatures reach 7 to 10°C.

Rhubarb Marmalade

Makes about 6 half-pint (250-mL) jars.

4 cups	rhubarb, chopped	1000 mL
2 cups	apples, peeled and diced	500 mL
1	orange	1
1	lemon	1
2 cups	brown sugar	500 mL
1 cup	raisins	250 mL
1 cup	currants	250 mL
½ cup	citron peel	125 mL
½ cup	apple juice	125 mL
½ tsp.	salt	2 mL
½ tsp.	nutmeg	2 mL
½ tsp.	allspice	2 mL
½ tsp.	cloves	2 mL
½ tsp.	cinnamon	2 mL

Place rhubarb and apple in large kettle. Grate the rind of the orange and lemon; add to rhubarb. Peel white membrane from orange and lemon and remove seeds. Cut into small pieces; stir into rhubarb. Mix remaining ingredients into rhubarb. Bring mixture to boil, reduce heat to low, and simmer 20 minutes or until mixture is thickened. Stir occasionally to make sure mixture does not stick to bottom. If not using immediately, place in jelly jars and process in water bath for 15 minutes. Use in tarts, pies, or cookies as you would mincemeat. Good stirred into plain yogurt.

condiments & sauces

more than just pies

Growing Rhubarb

Rhubarb plants are cool-season perennials and one of the first plants to be ready for harvesting in the spring. Rhubarb does not thrive at temperatures above 25°C, and quality fades as the summer days become hotter. Rhubarb will grow in a wide range of soils, as long as they are rich and well drained. Plants prefer slightly acidic soils and are heavy consumers of mineral nutrients; fertilizer applications of 110 gm/m^{-2} are recommended using a nitrogen fertilizer with 21% nitrogen content.

Early horticulturists had an interesting learning experience with rhubarb. It took some time and experimentation to determine the best combination of growing conditions. They learned in the eighteenth century that rhubarb responds well to heavy dunging, and in the nineteenth century that it needs a period of winter dormancy provided by a freezing climate.

Rhubarb is normally propagated by planting "sets," each comprising a fleshy rootstock with at least one bud. Sets are planted in the autumn or early spring. On light soils, plant the sets so that 2.5 centimetres of soil covers the buds; on heavy or wet soils, plant the sets with buds just above the soil surface. Space the plants about 90 centimetres apart. Rhubarb plants will respond well to heavy mulching and a consistent supply of water, particularly in dry weather and when sets are becoming established. Rhubarb plants will benefit from a heavy dressing of compost or manure every autumn or spring, and a nitrogenous dressing or organic liquid feed in the spring. For best results, the flowering stems should be removed before they mature, as they will weaken the plant.

Rhubarb plants can be grown from seed; however, resulting plants may not be uniform or true to the parent plant. Seeds are best sown in situ and seedlings pricked out the following year. That plants did not always come true from seed has been known for centuries—there was much comment on "bastard" or "mule" plants. As cultivation of rhubarb for culinary purposes intensified and a large number of people vied for the most attractive petioles, it become routine to reproduce offsets to obtain desired attributes in next year's crop.

Rhubarb is harvested by cutting the leaf petioles in the late spring and summer. With plants raised from sets, light harvesting may begin during the first year after planting. With plants grown from seed, harvesting will have to wait until the second year. In subsequent years, one can harvest heavily until the quality starts to deteriorate in the summer.

In the early spring of 1815, a bed of rhubarb growing in the Chelsea Physic Garden was inadvertently covered with leaf mould that had been thrown up by ditch-digging. When the rhubarb emerged through this thick mulch, the stalks had been blanched. Trial at the table revealed two desirable qualities: improved appearance and flavour, and a saving in the quantity of sugar. Although this discovery was first publicly reported by Thomas Hare, assistant secretary of the Horticultural Society of London, it was immediately seized upon by other gardeners and horticulturist. Then in 1818, both Daniel Judd, gardener to Charles Campbell of Edmonton, north London, and Thomas A. Knight, president of the London Horticultural Society, submitted papers on techniques of forcing rhubarb.

To obtain a really early crop of rhubarb with beautifully red petioles, one can force the plants. The Vegetable Garden Displayed*—with three hundred photographs, published by the Royal Horticultural Society in 1947—gives details on how this is done. Clear the ground around the selected rhubarb crown or crowns. Place a bottomless barrel, or any large container, over each crown and put a false top on the barrel or container; place manure with high straw content around the barrel. The rhubarb petioles growing in the dark will be devoid of chlorophyll and will be a cheery red colour.*

Microwave Rhubarb-Orange Sauce

Makes 4 servings.

4 cups	rhubarb, in ½-inch (12-mm) pieces	1 L
1 cup	sugar	250 mL
¼ tsp.	cinnamon, ground	1 mL
¼ cup	orange or lemon juice	60 mL

Combine ingredients in microwave-safe container and cover. Cook in the microwave on high for 8 minutes, stirring twice, until rhubarb is tender.

Simple Sauce

Makes 4 cups (1 L).

4 cups	rhubarb, in 1-inch (2.5-cm) pieces	1 L
¾ cup	sugar	180 mL
	water	

Combine rhubarb and sugar in saucepan. Add water to cover bottom of pan. Steam slowly until rhubarb is soft.

Rhubarb Conserve with Nuts

Makes 4 half-pint (250-mL) jars.

4 cups	rhubarb	1 L
4 cups	sugar	1 kg
2 tbsp.	orange zest	30 mL
¾ cup	orange juice	180 mL
¼ tsp.	salt	1 mL
½ cup	chopped hazelnuts	125 mL

Cut rhubarb into ½-inch (6-mm) pieces. Combine rhubarb, sugar, orange zest and juice, and salt in a heavy pot. Stir over low heat until sugar is dissolved, then boil rapidly until thickened, stirring frequently, about 20 minutes. Stir in nuts.

Fruity Rhubarb Conserve

Makes 10 half-pint (250-mL) jars.

4 cups	rhubarb, chopped	1 L
5½ cups	sugar	1.4 kg
2	oranges	2
1	lemon	1
1½ cups	chopped dates **or** raisins	375 mL
½ cup	crystallized ginger	125 mL
1 cup	chopped walnuts	250 mL

Place rhubarb in a large pot; stir in sugar until well blended. Cover and let stand at room temperature overnight. Cut unpeeled oranges and lemon into thin slices; remove seeds and cut each slice into small pieces. Add to rhubarb-sugar mixture along with dates and ginger. Bring to a boil; reduce heat and simmer, uncovered, stirring occasionally, for 35 to 40 minutes or until thickened. About 5 minutes before removing from heat, stir in walnuts. Fill prepared jars to within ⅛ inch (3 mm) of rim. Process in hot water bath for 15 minutes. Store in a cool, dark place. Wonderful spread for crackers or toast, and an excellent accompaniment to roasted beef and lamb.

Rhubarb Syrup

Makes about 2½ cups (625 mL).

8 cups	rhubarb, coarsely chopped	2 L
1 cup	water	250 mL
2 cups	sugar	500 mL
1 tbsp.	cornstarch	15 mL
¼ cup	water	60 mL

Bring rhubarb to a boil in water. Simmer 5 to 10 minutes, until tender. Sieve though 4 layers of dampened cheesecloth. This should result in about 2 cups (500 mL) of juice. Measure, then return to saucepan with an equal amount of sugar. Bring to a boil. Dissolve cornstarch in ¼ cup (60 mL) water; stir into boiling syrup. Boil about 5 minutes or until slightly thickened. Seal in sterilized jars or store in refrigerator for up to 2 months. Good on pancakes or over ice cream.

Rhubarb-Lime Vinaigrette

Makes 1 cup (250 mL).

½ cup	rhubarb	125 mL
¼ cup	water	60 mL
1 tsp.	ginger root, finely chopped	5 mL
2 tbsp.	honey	30 mL
1	lime, juice and zest	1
½ cup	canola oil	125 mL
1 tbsp.	chives, minced	15 mL
1 tbsp.	fresh thyme, minced	15 mL
	salt, to taste	
	hot chili flakes, to taste	

Chop rhubarb and simmer with water in a small, covered pot until tender. Stir in the ginger and honey while the fruit is still warm. Add lime juice and zest, oil, chives, and thyme; whisk, seasoning to taste with salt and chili flakes. Excellent served over mixed greens with grilled salmon.

Rhubarb Chutney

Makes 1 cup (250 mL).

4 cups	rhubarb	1 L
2 tsp.	fresh ginger, grated	10 mL
2	garlic cloves	2
1 or 2	jalapeño peppers	1 or 2
1 tsp.	paprika	5 mL
1 tbsp.	black mustard seeds	15 mL
¼ cup	dried currants	60 mL
1 cup	light brown sugar	250 mL
1½ cups	white-wine vinegar	375 mL

Wash rhubarb and cut into ¼-inch (6-mm) pieces. Seed and remove veins from jalapeños; chop the jalapeños and garlic finely. Place all ingredients in a saucepan; bring to a boil. Lower heat and simmer until rhubarb breaks down and mixture is the texture of jam, about 45 minutes. Keeps for several months stored in a glass jar in the refrigerator.

Rich Rhubarb Chutney

Makes about 4 cups (1 L).

8 cups	rhubarb, diced	2 L
3 tbsp.	sugar	45 mL
½ cup	honey	125 mL
½ cup	dried cherries*	125 mL
5 tbsp.	red-wine vinegar	75 mL
3 tbsp.	dry red wine	45 mL
2 tsp.	mustard seeds	10 mL
½ tsp.	kosher salt	2 mL
pinch	cinnamon	pinch
pinch	allspice	pinch
pinch	cayenne	pinch
½ cup	red onion, minced	125 mL
¾ cup	celery	180 mL
1 tbsp.	jalapeño, seeded and minced	15 mL
2 tsp.	orange zest	10 mL
2 tsp.	lime juice	10 mL

Place rhubarb in a colander, sprinkle with sugar, and drain for 30 minutes. Cook the honey, cherries, vinegar, wine, mustard seeds, salt, cinnamon, allspice, and cayenne in a 10-inch (25-cm) skillet over medium heat, stirring occasionally, until syrupy. Mince onion and cut celery into ½-inch (12-mm) slices. Stir into cherry mixture and cook for 2 minutes. Add the rhubarb and cook over medium heat for 10 minutes, stirring occasionally. Avoid overcooking (the rhubarb should be tender but not mushy). Add the jalapeño and orange zest; cook 1 minute. Stir in the lime juice, remove from the pan, and cool. Serve at room temperature or chilled.

**Dried raisins, dried cranberries, or minced dried apricots may be substituted for dried cherries.*

Pie Plant Chutney

Makes 6 cups (1.5 L)

8 cups	rhubarb	2 L
1	red bell pepper	1
1	orange	1
½	lemon	1/2
1½ cups	chopped onion	375 mL
½ cup	golden raisins	125 mL
1	clove garlic, minced	1
1 cup	sugar	250 mL
1 cup	brown sugar	250 mL
1 cup	vinegar	250 mL
½ tsp.	cinnamon	2 mL
¼ tsp.	ground cloves	1 mL
¼ tsp.	cayenne pepper	1 mL
½ tsp.	salt	2 mL

Prepare rhubarb by chopping stalks into ½-inch (12-mm) pieces. Seed and coarsely chop red pepper. Remove seeds from orange and lemon and chop finely (include rind). Combine all ingredients in a large stainless-steel saucepan. Cover and bring to a boil. Simmer gently for one hour or until mixture is thick enough to mound in a spoon. The chutney will keep for up to 4 weeks in the refrigerator, or may be canned by water-bath method, processing 10 minutes.

Saskatoon-Rhubarb Chutney

Makes 12 cups (3 L).

8 cups	rhubarb, chopped	2 L
8 cups	onions, chopped	2 L
4 cups	saskatoon berries	1 L
3 cups	cider vinegar	750 mL
1 cup	raisins	250 mL
1 cup	sugar	250 mL
1/4 cup	candied ginger, minced	60 mL
2 tsp.	ground cloves	10 mL
2 tsp.	cinnamon	10 mL
2 tsp.	salt	10 mL
1 tsp.	pepper	5 mL

Combine all ingredients in a Dutch oven; bring to a boil. Reduce heat and simmer for 30 minutes or until thickened, stirring frequently. Ladle into canning jars; seal and process in boiling water bath for 10 minutes.

Rhubarb Chutney with Dates

Makes about 10 cups (2.5 L).

8 cups	rhubarb, chopped	2 L
1 lb.	stoned dates	454 g
5 cups	brown sugar	1.25 kg
1 tbsp.	ground cinnamon	15 mL
1 tbsp.	ground cloves	15 mL
2 cups	cider vinegar	500 mL

Combine rhubarb, dates, brown sugar, cinnamon, and cloves in large pot and boil slowly for 2 hours. Stir frequently to avoid scorching. Add the vinegar and boil 10 minutes. The chutney will keep well in the refrigerator or you may process. To process, pack the chutney into clean, hot pint jars, leaving 1/2-inch (12-mm) headspace. Seal. Process in a boiling water bath for 10 minutes.

Rhubarb-Carrot Relish

Makes about 3 cups (750 mL).

	salt	
1	large carrot	1
6 cups	rhubarb, diced	1.5 L
2 tsp.	mustard seeds	10 mL
1 tbsp.	softened butter **or** walnut oil	15 mL

Set 2 pans of salted water on the stove and bring both to a boil. Peel and cut carrot into chunks. Cook the carrot pieces in salted water for 15 to 20 minutes, or until soft. Cook the rhubarb in the other pan of salted water for 3 to 5 minutes, or until just tender. When the rhubarb is done, drain and crush it with a fork or potato masher. When the carrot chunks are done, drain and purée them. Combine the rhubarb, carrot, mustard seeds, and butter or oil. Serve at room temperature with steamed fish, grilled chicken, or other meat.

Rhubarb Relish

Makes about 3 cups (750 mL).

4 cups	rhubarb, diced	1 L
4 lbs.	brown sugar	1.8 kg
2 cups	vinegar	500 mL
2 tsp.	salt	10 mL
1	onion, chopped	1
1 tsp.	cinnamon	5 mL
1 tsp.	allspice	5 mL
½ tsp.	cloves	2 mL
2 tsp.	salt	10 mL
½ tsp.	pepper	2 mL

Combine all ingredients in large saucepan and boil until thickened. Place in jars and seal. Very good with steak and beef roasts.

main dishes & salads

more than. just pies

Other Uses for Rhubarb

Rhubarb is 95% water and somewhat acidic (in most recipes this is offset by the addition of sugar), but it is a good source of potassium and contributes small amounts of vitamins (with the greatest concentration being vitamin C). It is low in sodium and is a good source of fibre. One cup of rhubarb contains about 26 calories.

The use of rhubarb as a fruit (or by some as a vegetable) was an offshoot of the great popularity of medicinal rhubarb in Europe, most especially in Britain. It was consumed as food before the nineteenth century but not extensively. In Elizabethan times, the leaves were reportedly used as a table green rather in the manner of spinach or beet root leaves. At the time, it was not known that the oxalic acid in these leaves rendered them toxic, even to the point of death. Some professed that the leaves were palatable but most reacted like Forsyth of Alton towers: "I tasted some boiled, and they did not appear to me to have one redeeming quality to keep them an instant from the dung heap" (Gardeners Chronicle, *1847, 20: 325).*

It was during the eighteenth century, when sugar became increasingly available as well as steadily cheaper, that changes in English and Continental diets occurred. In a letter dated 2nd September 1739 to the gardener John Bartram of London, Peter Collinson recommended that Bartram experiment with R. undulatum *(Siberian rhubarb), which he insisted would make a good tart:*

"All you have to do, is to take the stalks from the root, and from the leaves; peel off the rind and cut them into 2 or 3 pieces, and put them in a crust with sugar and a little cinnamon; then bake the pie or tart; eats best when cold. It is much admired here, and none has the effect that roots have. It eats like gooseberry pie."

One of the most intriguing uses was the preparation of the flowers in a manner similar to broccoli or cauliflower. James Barnes of Bicton, near Shrewsbury in Shropshire, found the inflorescences superior to the petioles. However, this use did not catch on.

Rhubarb has also been used for cleaning cooking utensils and as an organic insecticide to kill leaf-eating insects. Even a fairly strong dye used for colouring hair can be made from rhubarb.

Pork Chops and Rhubarb Casserole

Makes 6 servings.

6	pork chops, 1 inch (25 mm) thick	6
1 tbsp.	oil	15 mL
1 tbsp.	butter	15 mL
	salt, to taste	
	fresh ground pepper, to taste	
2 cups	fine, fresh bread crumbs	500 mL
½ cup	sugar	125 mL
½ cup	brown sugar	125 mL
3 tbsp.	flour	45 mL
½ tsp.	cinnamon	2 mL
¼ tsp.	salt	1 mL
6 cups	rhubarb, thinly sliced	1.5 L

Preheat oven to 350°F (180°C). Grease a shallow casserole. Trim fat from chops. In a large, heavy skillet, quickly brown chops in oil and butter, and season with salt and pepper. Set them aside, then pour any pan drippings over bread crumbs, mixing in with a fork. Combine sugars, flour, cinnamon, salt, and rhubarb. Sprinkle bottom of casserole with half the bread crumbs, then cover with half of rhubarb mixture. Place chops on top, then cover with the rest of the rhubarb mixture (but not remaining crumbs). Cover casserole tightly and bake 40 minutes. Remove cover, top with rest of bread crumbs, and bake another 10 minutes.

Grilled Pork Chops with Asparagus Salad and Rhubarb Sauce

Makes 4 servings.

4 stalks	rhubarb	4 stalks
1 cup	red wine or port	250 mL
½ cup	red-wine vinegar	125 mL
¾ cup	chicken stock or water	180 mL
1 lb.	asparagus	454 g
1	orange, zest and juice	1

2 tbsp.	fresh parsley, fresh	30 mL
½ cup	olive oil	125 mL
	salt and pepper, to taste	
4 cups	spinach leaves, loosely packed	1 L
4	boneless pork chops	4

Combine rhubarb, red wine or port, and vinegar in a medium-sized stainless-steel saucepan. Let rhubarb marinate for 30 minutes. Add stock or water to saucepan and bring to a slow boil. Cook until rhubarb falls apart and liquid is thick enough to coat a spoon, about 25 to 30 minutes. Keep warm. Prepare grill or broiler. Cut asparagus into 2-inch (5-cm) lengths and cook until just tender. Cool in cold water and set aside. Whisk together the orange juice, zest, and parsley. While whisking, slowly add the olive oil. Season with salt and pepper and toss with the asparagus and spinach. Set aside. Season the pork chops with salt and pepper. Grill or broil until cooked through, about 8 minutes, depending on thickness. Arrange the asparagus and spinach salad on 4 serving plates. Place the pork chops on top of the salad. Spoon the rhubarb sauce over the pork and serve.

Lamb Khoresh with Rhubarb

Makes 6 to 8 servings.

2	large onions, chopped	2
2 tbsp.	butter	30 mL
3 lbs.	boned shoulder of lamb	1.4 kg
pinch	saffron	pinch
2½ cups	beef stock	625 mL
4 tbsp.	lemon juice	60 mL
	salt and pepper, to taste	
2	bunches fresh parsley	2
6	sprigs fresh mint	6
2 tbsp.	butter	30 mL
4 cups	rhubarb	1 L

In a large pot, brown onions in butter. Push to one side and brown the meat. Add the saffron and stir. Add the stock, lemon juice, salt, and pepper; cover and simmer for 1 hour. Chop the parsley with the mint and fry in butter in a small frypan. Add to the stew and simmer another ½ hour. Cut the rhubarb into 1-inch (25-mm) pieces and add to the stew for the last 15 minutes. Remove the meat with a slotted spoon to a serving dish. Skim the fat from the liquid and boil the liquid hard to reduce by a third; pour over the meat. Serve with basmati rice.

Rhubarb Curry

Makes 4 servings.

2	celery stalks	2
2	potatoes	2
1	sweet potato	1
1	green banana	1
1	2-inch (5-cm) piece of ginger	1
2	carrots	2
2 tbsp.	olive oil	30 mL
½ tsp.	mustard seeds	2 mL
2 cups	rhubarb purée	500 mL
1 tsp.	salt	5 mL
1 tsp.	sugar	5 mL
2 cups	water	500 mL

To prepare vegetables, cut celery into ½-inch (12-mm) slices; cube potato and sweet potato into ½-inch (12-mm) pieces. Cut banana into 1-inch (25-mm) rounds. Grate peeled ginger finely. Heat the oil in a large saucepan over medium heat for 2 minutes. Add the mustard seeds; when they begin to pop, reduce heat, add the ginger, and fry for 30 seconds. Add the vegetables, salt, and sugar; stirfry for 5 minutes. Add water and rhubarb purée, bring to a boil, lower the heat, cover, and cook for 10 minutes.

Rhubarb Festive Spring Salad

Makes 4 servings.

2 cups	rhubarb, cut into 1½-inch (37-mm) pieces	500 mL
½ cup	sugar	125 mL
½ cup	water	125 mL
1 package	apple-flavoured gelatin	1 package
3 oz.	cream cheese	90 g
1 tbsp.	lemon juice	30 mL
⅔ cup	sliced celery	160 mL
⅓ cup	chopped nuts	80 mL

Cook rhubarb with sugar and water. Bring to boil and reduce heat to low for 10 minutes. Dissolve the gelatin in the boiling hot rhubarb sauce. Add the cream cheese, which has been mashed to break it up, and stir until dissolved in the hot mixture. Chill until it begins to thicken. Whip until light and fluffy, then add lemon juice, celery, and nuts. Turn into a salad mould or individual moulds. Chill.

Celina's Rhubarb Curry

Makes 12 servings.

1	eggplant, large	1
½	small pumpkin/squash	½
1	large Asian radish (daikon)	1
6	carrots	6
4	large potatoes	4
4	large tomatoes	4
1 cup	toovar dal	250 mL
1 cup	water	250 mL
	tamarind pulp, ½ lime-size ball	
1 cup	rhubarb	250 mL
½ cup	warm water	125 mL
1 cup	corn/peanut oil	250 mL
2	onions, chopped	2
1	2-inch (5-cm) piece of ginger	1
½ tsp.	ground turmeric	2 mL
	salt	
2 tsp.	ground cumin	10 mL
1 tsp.	ground coriander	5 mL
1 tsp.	paprika	5 mL
4	bay leaves	4
4 cups	water	1 L
1 can	coconut milk	1 can
½ cup	cilantro leaves	125 mL

Cut eggplant, pumpkin, daikon, and carrots into 1-inch (25-mm) cubes. Dice potatoes and tomatoes. In a saucepan, combine the toovar dal and 1 cup (250 mL) water; boil for 20 minutes. Set aside. Combine the tamarind, rhubarb, and water in a saucepan and bring to a boil. Lower heat and simmer for 10 minutes; strain and discard the pulp. Save the liquid (note: 1 teaspoon [5 mL] lime juice can be used for the tamarind). In a large skillet, heat the oil over medium heat. Add the onions and sauté for 2 minutes. Add the ginger, turmeric, salt, cumin, coriander, paprika, and bay leaves, and sauté for a further 2 minutes. Add the toovar dal and water from the saucepan. Cover and cook for 2 minutes. Add the tamarind liquid, vegetables, water, and coconut milk; cover and simmer for 20 minutes. Garnish with the cilantro leaves just before serving.

Further Reading

Coats, A.M. *The Plant Hunters*. New York: McGraw-Hill Book Company, 1970.

Demars, J. *Rhubarb Recipes*. Apple Blossom Books, 1994.

Foust, C. M. *Rhubarb: The Wondrous Drug*. New Jersey: Princeton University Press, 1992.

Macqueen Cowan, J., ed. *The Journeys and Plant Introductions of George Forrest V.M.H.* Oxford University Press, 1952.

The New Royal Horticultural Society Dictionary—Index of Garden Plants. Portland, Oregon: Timber Press, 1994.

The New Royal Horticultural Society Dictionary of Gardening. Volumes I to VI. London and Basingstoke: The Macmillan Press, 1994.

The Rhubarb Compendium. Http://www.clark.net/pub/dan/rhubarb.html.

"Rhubarb: Culture and Physiology." *Chemistry and Technology*. Ontario Department of Agriculture and Food, 1967.

The Royal Horticultural Society A–Z Encyclopedia of Garden Plants. Volumes I and II. London: Dorling Kindersley, 1996.

Sondra, A.S., ed. *Exclusively Rhubarb Cookbook*. Coventry, Connecticut: Coventry Historical Society, 1992.

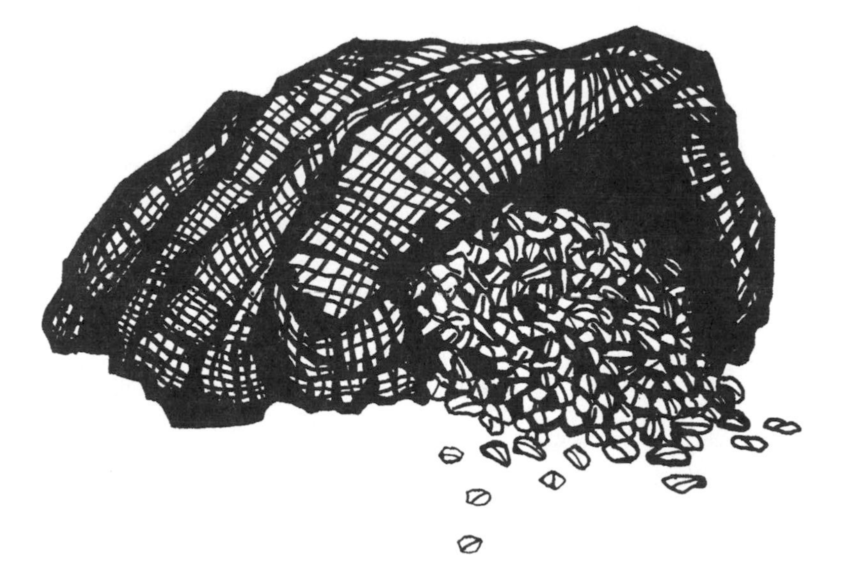

Imperial to Metric Conversion

Oven Temperatures

250°F = 120°C

275°F = 140°C

300°F = 150°C

325°F = 160°C

350°F = 180°C

375°F = 190°C

400°F = 200°C

425°F = 220°C

450°F = 230°C

475°F = 250°C

500°F = 260°C

525°F = 270°C

Volume

Pinch ≤ 1/16 tsp.

1/16 tsp. = 0.25 mL

1/8 tsp. = 0.5 mL

1/4 tsp. = 1 mL

1/2 tsp. = 2 mL

1 tsp. = 5 mL

1 tbsp. = 15 mL

2 tbsp. = 30 mL

3 tbsp. = 45 mL

1/4 cup = 60 mL

1/3 cup = 80 mL

1/2 cup = 125 mL

2/3 cup = 160 mL

3/4 cup = 180 mL

1 cup = 250 mL

Length

1 inch = 25 mm

Weight

1 oz. = 30 grams